Pitman Research Notes in Mathematics Series

AF699170

Submission of proposals for consideration

Suggestions for publication, in the form of outlines and representative samples, are invited by the Editorial Board for assessment. Intending authors should approach one of the main editors or another member of the Editorial Board, citing the relevant AMS subject classifications. Alternatively, outlines may be sent directly to the publisher's offices. Refereeing is by members of the board and other mathematical authorities in the topic concerned, throughout the world.

Preparation of accepted manuscripts

On acceptance of a proposal, the publisher will supply full instructions for the preparation of manuscripts in a form suitable for direct photo-lithographic reproduction. Specially printed grid sheets can be provided and a contribution is offered by the publisher towards the cost of typing. Word processor output, subject to the publisher's approval, is also acceptable.

Illustrations should be prepared by the authors, ready for direct reproduction without further improvement. The use of hand-drawn symbols should be avoided wherever possible, in order to maintain maximum clarity of the text.

The publisher will be pleased to give any guidance necessary during the preparation of a typescript, and will be happy to answer any queries.

Important note

In order to avoid later retyping, intending authors are strongly urged not to begin final preparation of a typescript before receiving the publisher's guidelines. In this way it is hoped to preserve the uniform appearance of the series.

Longman Group Ltd
Longman House
Burnt Mill
Harlow, Essex, CM20 2JE
UK
(Telephone (0) 1279 426721)

Titles in this series. A full list is available from the publisher on request.

100 Optimal control of variational inequalities
V Barbu
101 Partial differential equations and dynamical systems
W E Fitzgibbon III
102 Approximation of Hilbert space operators Volume II
C Apostol, L A Fialkow, D A Herrero and D Voiculescu
103 Nondiscrete induction and iterative processes
V Ptak and F-A Potra
104 Analytic functions – growth aspects
O P Juneja and G P Kapoor
105 Theory of Tikhonov regularization for Fredholm equations of the first kind
C W Groetsch
106 Nonlinear partial differential equations and free boundaries. Volume I
J I Díaz
107 Tight and taut immersions of manifolds
T E Cecil and P J Ryan
108 A layering method for viscous, incompressible L_p flows occupying R^n
A Douglis and E B Fabes
109 Nonlinear partial differential equations and their applications: Collège de France Seminar. Volume VI
H Brezis and J L Lions
110 Finite generalized quadrangles
S E Payne and J A Thas
111 Advances in nonlinear waves. Volume II
L Debnath
112 Topics in several complex variables
E Ramírez de Arellano and D Sundararaman
113 Differential equations, flow invariance and applications
N H Pavel
114 Geometrical combinatorics
F C Holroyd and R J Wilson
115 Generators of strongly continuous semigroups
J A van Casteren
116 Growth of algebras and Gelfand–Kirillov dimension
G R Krause and T H Lenagan
117 Theory of bases and cones
P K Kamthan and M Gupta
118 Linear groups and permutations
A R Camina and E A Whelan
119 General Wiener–Hopf factorization methods
F-O Speck
120 Free boundary problems: applications and theory. Volume III
A Bossavit, A Damlamian and M Fremond
121 Free boundary problems: applications and theory. Volume IV
A Bossavit, A Damlamian and M Fremond
122 Nonlinear partial differential equations and their applications: Collège de France Seminar. Volume VII
H Brezis and J L Lions
123 Geometric methods in operator algebras
H Araki and E G Effros
124 Infinite dimensional analysis–stochastic processes
S Albeverio
125 Ennio de Giorgi Colloquium
P Krée
126 Almost-periodic functions in abstract spaces
S Zaidman
127 Nonlinear variational problems
A Marino, L Modica, S Spagnolo and M Degliovanni
128 Second-order systems of partial differential equations in the plane
L K Hua, W Lin and C-Q Wu
129 Asymptotics of high-order ordinary differential equations
R B Paris and A D Wood
130 Stochastic differential equations
R Wu
131 Differential geometry
L A Cordero
132 Nonlinear differential equations
J K Hale and P Martinez-Amores
133 Approximation theory and applications
S P Singh
134 Near-rings and their links with groups
J D P Meldrum
135 Estimating eigenvalues with *a posteriori/a priori* inequalities
J R Kuttler and V G Sigillito
136 Regular semigroups as extensions
F J Pastijn and M Petrich
137 Representations of rank one Lie groups
D H Collingwood
138 Fractional calculus
G F Roach and A C McBride
139 Hamilton's principle in continuum mechanics
A Bedford
140 Numerical analysis
D F Griffiths and G A Watson
141 Semigroups, theory and applications. Volume I
H Brezis, M G Crandall and F Kappel
142 Distribution theorems of L-functions
D Joyner
143 Recent developments in structured continua
D De Kee and P Kaloni
144 Functional analysis and two-point differential operators
J Locker
145 Numerical methods for partial differential equations
S I Hariharan and T H Moulden
146 Completely bounded maps and dilations
V I Paulsen
147 Harmonic analysis on the Heisenberg nilpotent Lie group
W Schempp
148 Contributions to modern calculus of variations
L Cesari
149 Nonlinear parabolic equations: qualitative properties of solutions
L Boccardo and A Tesei
150 From local times to global geometry, control and physics
K D Elworthy

151 A stochastic maximum principle for optimal control of diffusions
U G Haussmann
152 Semigroups, theory and applications. Volume II
H Brezis, M G Crandall and F Kappel
153 A general theory of integration in function spaces
P Muldowney
154 Oakland Conference on partial differential equations and applied mathematics
L R Bragg and J W Dettman
155 Contributions to nonlinear partial differential equations. Volume II
J I Díaz and P L Lions
156 Semigroups of linear operators: an introduction
A C McBride
157 Ordinary and partial differential equations
B D Sleeman and R J Jarvis
158 Hyperbolic equations
F Colombini and M K V Murthy
159 Linear topologies on a ring: an overview
J S Golan
160 Dynamical systems and bifurcation theory
M I Camacho, M J Pacifico and F Takens
161 Branched coverings and algebraic functions
M Namba
162 Perturbation bounds for matrix eigenvalues
R Bhatia
163 Defect minimization in operator equations: theory and applications
R Reemtsen
164 Multidimensional Brownian excursions and potential theory
K Burdzy
165 Viscosity solutions and optimal control
R J Elliott
166 Nonlinear partial differential equations and their applications: Collège de France Seminar. Volume VIII
H Brezis and J L Lions
167 Theory and applications of inverse problems
H Haario
168 Energy stability and convection
G P Galdi and B Straughan
169 Additive groups of rings. Volume II
S Feigelstock
170 Numerical analysis 1987
D F Griffiths and G A Watson
171 Surveys of some recent results in operator theory. Volume I
J B Conway and B B Morrel
172 Amenable Banach algebras
J-P Pier
173 Pseudo-orbits of contact forms
A Bahri
174 Poisson algebras and Poisson manifolds
K H Bhaskara and K Viswanath
175 Maximum principles and eigenvalue problems in partial differential equations
P W Schaefer
176 Mathematical analysis of nonlinear, dynamic processes
K U Grusa
177 Cordes' two-parameter spectral representation theory
D F McGhee and R H Picard
178 Equivariant K-theory for proper actions
N C Phillips
179 Elliptic operators, topology and asymptotic methods
J Roe
180 Nonlinear evolution equations
J K Engelbrecht, V E Fridman and E N Pelinovski
181 Nonlinear partial differential equations and their applications: Collège de France Seminar. Volume IX
H Brezis and J L Lions
182 Critical points at infinity in some variational problems
A Bahri
183 Recent developments in hyperbolic equations
L Cattabriga, F Colombini, M K V Murthy and S Spagnolo
184 Optimization and identification of systems governed by evolution equations on Banach space
N U Ahmed
185 Free boundary problems: theory and applications. Volume I
K H Hoffmann and J Sprekels
186 Free boundary problems: theory and applications. Volume II
K H Hoffmann and J Sprekels
187 An introduction to intersection homology theory
F Kirwan
188 Derivatives, nuclei and dimensions on the frame of torsion theories
J S Golan and H Simmons
189 Theory of reproducing kernels and its applications
S Saitoh
190 Volterra integrodifferential equations in Banach spaces and applications
G Da Prato and M Iannelli
191 Nest algebras
K R Davidson
192 Surveys of some recent results in operator theory. Volume II
J B Conway and B B Morrel
193 Nonlinear variational problems. Volume II
A Marino and M K V Murthy
194 Stochastic processes with multidimensional parameter
M E Dozzi
195 Prestressed bodies
D Iesan
196 Hilbert space approach to some classical transforms
R H Picard
197 Stochastic calculus in application
J R Norris
198 Radical theory
B J Gardner
199 The C^*-algebras of a class of solvable Lie groups
X Wang
200 Stochastic analysis, path integration and dynamics
K D Elworthy and J C Zambrini

201 Riemannian geometry and holonomy groups
S Salamon
202 Strong asymptotics for extremal errors and polynomials associated with Erdös type weights
D S Lubinsky
203 Optimal control of diffusion processes
V S Borkar
204 Rings, modules and radicals
B J Gardner
205 Two-parameter eigenvalue problems in ordinary differential equations
M Faierman
206 Distributions and analytic functions
R D Carmichael and D Mitrovic
207 Semicontinuity, relaxation and integral representation in the calculus of variations
G Buttazzo
208 Recent advances in nonlinear elliptic and parabolic problems
P Bénilan, M Chipot, L Evans and M Pierre
209 Model completions, ring representations and the topology of the Pierce sheaf
A Carson
210 Retarded dynamical systems
G Stepan
211 Function spaces, differential operators and nonlinear analysis
L Paivarinta
212 Analytic function theory of one complex variable
C C Yang, Y Komatu and K Niino
213 Elements of stability of visco-elastic fluids
J Dunwoody
214 Jordan decomposition of generalized vector measures
K D Schmidt
215 A mathematical analysis of bending of plates with transverse shear deformation
C Constanda
216 Ordinary and partial differential equations. Volume II
B D Sleeman and R J Jarvis
217 Hilbert modules over function algebras
R G Douglas and V I Paulsen
218 Graph colourings
R Wilson and R Nelson
219 Hardy-type inequalities
A Kufner and B Opic
220 Nonlinear partial differential equations and their applications: Collège de France Seminar. Volume X
H Brezis and J L Lions
221 Workshop on dynamical systems
E Shiels and Z Coelho
222 Geometry and analysis in nonlinear dynamics
H W Broer and F Takens
223 Fluid dynamical aspects of combustion theory
M Onofri and A Tesei
224 Approximation of Hilbert space operators. Volume I. 2nd edition
D Herrero
225 Operator theory: proceedings of the 1988 GPOTS–Wabash conference
J B Conway and B B Morrel
226 Local cohomology and localization
J L Bueso Montero, B Torrecillas Jover and A Verschoren
227 Nonlinear waves and dissipative effects
D Fusco and A Jeffrey
228 Numerical analysis 1989
D F Griffiths and G A Watson
229 Recent developments in structured continua. Volume II
D De Kee and P Kaloni
230 Boolean methods in interpolation and approximation
F J Delvos and W Schempp
231 Further advances in twistor theory. Volume I
L J Mason and L P Hughston
232 Further advances in twistor theory. Volume II
L J Mason, L P Hughston and P Z Kobak
233 Geometry in the neighborhood of invariant manifolds of maps and flows and linearization
U Kirchgraber and K Palmer
234 Quantales and their applications
K I Rosenthal
235 Integral equations and inverse problems
V Petkov and R Lazarov
236 Pseudo-differential operators
S R Simanca
237 A functional analytic approach to statistical experiments
I M Bomze
238 Quantum mechanics, algebras and distributions
D Dubin and M Hennings
239 Hamilton flows and evolution semigroups
J Gzyl
240 Topics in controlled Markov chains
V S Borkar
241 Invariant manifold theory for hydrodynamic transition
S Sritharan
242 Lectures on the spectrum of $L^2(\Gamma\backslash G)$
F L Williams
243 Progress in variational methods in Hamiltonian systems and elliptic equations
M Girardi, M Matzeu and F Pacella
244 Optimization and nonlinear analysis
A Ioffe, M Marcus and S Reich
245 Inverse problems and imaging
G F Roach
246 Semigroup theory with applications to systems and control
N U Ahmed
247 Periodic-parabolic boundary value problems and positivity
P Hess
248 Distributions and pseudo-differential operators
S Zaidman
249 Progress in partial differential equations: the Metz surveys
M Chipot and J Saint Jean Paulin
250 Differential equations and control theory
V Barbu

251 Stability of stochastic differential equations with respect to semimartingales
X Mao
252 Fixed point theory and applications
J Baillon and M Théra
253 Nonlinear hyperbolic equations and field theory
M K V Murthy and S Spagnolo
254 Ordinary and partial differential equations. Volume III
B D Sleeman and R J Jarvis
255 Harmonic maps into homogeneous spaces
M Black
256 Boundary value and initial value problems in complex analysis: studies in complex analysis and its applications to PDEs 1
R Kühnau and W Tutschke
257 Geometric function theory and applications of complex analysis in mechanics: studies in complex analysis and its applications to PDEs 2
R Kühnau and W Tutschke
258 The development of statistics: recent contributions from China
X R Chen, K T Fang and C C Yang
259 Multiplication of distributions and applications to partial differential equations
M Oberguggenberger
260 Numerical analysis 1991
D F Griffiths and G A Watson
261 Schur's algorithm and several applications
M Bakonyi and T Constantinescu
262 Partial differential equations with complex analysis
H Begehr and A Jeffrey
263 Partial differential equations with real analysis
H Begehr and A Jeffrey
264 Solvability and bifurcations of nonlinear equations
P Drábek
265 Orientational averaging in mechanics of solids
A Lagzdins, V Tamuzs, G Teters and A Kregers
266 Progress in partial differential equations: elliptic and parabolic problems
C Bandle, J Bemelmans, M Chipot, M Grüter and J Saint Jean Paulin
267 Progress in partial differential equations: calculus of variations, applications
C Bandle, J Bemelmans, M Chipot, M Grüter and J Saint Jean Paulin
268 Stochastic partial differential equations and applications
G Da Prato and L Tubaro
269 Partial differential equations and related subjects
M Miranda
270 Operator algebras and topology
W B Arveson, A S Mishchenko, M Putinar, M A Rieffel and S Stratila
271 Operator algebras and operator theory
W B Arveson, A S Mishchenko, M Putinar, M A Rieffel and S Stratila
272 Ordinary and delay differential equations
J Wiener and J K Hale
273 Partial differential equations
J Wiener and J K Hale
274 Mathematical topics in fluid mechanics
J F Rodrigues and A Sequeira
275 Green functions for second order parabolic integro-differential problems
M G Garroni and J F Menaldi
276 Riemann waves and their applications
M W Kalinowski
277 Banach C(K)-modules and operators preserving disjointness
Y A Abramovich, E L Arenson and A K Kitover
278 Limit algebras: an introduction to subalgebras of C*-algebras
S C Power
279 Abstract evolution equations, periodic problems and applications
D Daners and P Koch Medina
280 Emerging applications in free boundary problems
J Chadam and H Rasmussen
281 Free boundary problems involving solids
J Chadam and H Rasmussen
282 Free boundary problems in fluid flow with applications
J Chadam and H Rasmussen
283 Asymptotic problems in probability theory: stochastic models and diffusions on fractals
K D Elworthy and N Ikeda
284 Asymptotic problems in probability theory: Wiener functionals and asymptotics
K D Elworthy and N Ikeda
285 Dynamical systems
R Bamon, R Labarca, J Lewowicz and J Palis
286 Models of hysteresis
A Visintin
287 Moments in probability and approximation theory
G A Anastassiou
288 Mathematical aspects of penetrative convection
B Straughan
289 Ordinary and partial differential equations. Volume IV
B D Sleeman and R J Jarvis
290 *K*-theory for real *C**-algebras
H Schröder
291 Recent developments in theoretical fluid mechanics
G P Galdi and J Necas
292 Propagation of a curved shock and nonlinear ray theory
P Prasad
293 Non-classical elastic solids
M Ciarletta and D Ieşan
294 Multigrid methods
J Bramble
295 Entropy and partial differential equations
W A Day
296 Progress in partial differential equations: the Metz surveys 2
M Chipot
297 Nonstandard methods in the calculus of variations
C Tuckey
298 Barrelledness, Baire-like- and (LF)-spaces
M Kunzinger
299 Nonlinear partial differential equations and their applications. Collège de France Seminar. Volume XI
H Brezis and J L Lions
300 Introduction to operator theory
T Yoshino

301 Generalized fractional calculus and applications
V Kiryakova
302 Nonlinear partial differential equations and their applications. Collège de France Seminar Volume XII
H Brezis and J L Lions
303 Numerical analysis 1993
D F Griffiths and G A Watson
304 Topics in abstract differential equations
S Zaidman
305 Complex analysis and its applications
C C Yang, G C Wen, K Y Li and Y M Chiang
306 Computational methods for fluid-structure interaction
J M Crolet and R Ohayon
307 Random geometrically graph directed self-similar multifractals
L Olsen
308 Progress in theoretical and computational fluid mechanics
G P Galdi, J Málek and J Necas
309 Variational methods in Lorentzian geometry
A Masiello
310 Stochastic analysis on infinite dimensional spaces
H Kunita and H-H Kuo
311 Representations of Lie groups and quantum groups
V Baldoni and M Picardello
312 Common zeros of polynomials in several variables and higher dimensional quadrature
Y Xu
313 Extending modules
N V Dung, D van Huynh, P F Smith and R Wisbauer
314 Progress in partial differential equations: the Metz surveys 3
M Chipot, J Saint Jean Paulin and I Shafrir
315 Refined large deviation limit theorems
V Vinogradov
316 Topological vector spaces, algebras and related areas
A Lau and I Tweddle
317 Integral methods in science and engineering
C Constanda
318 A method for computing unsteady flows in porous media
R Raghavan and E Ozkan
319 Asymptotic theories for plates and shells
R P Gilbert and K Hackl
320 Nonlinear variational problems and partial differential equations
A Marino and M K V Murthy
321 Topics in abstract differential equations II
S Zaidman
322 Diffraction by wedges
B Budaev
323 Free boundary problems: theory and applications
J I Diaz, M A Herrero, A Liñan and J L Vazquez
324 Recent developments in evolution equations
A C McBride and G F Roach
325 Elliptic and parabolic problems: Pont-à-Mousson 1994
C Bandle, J Bemelmans, M Chipot, J Saint Jean Paulin and I Shafrir
326 Calculus of variations, applications and computations: Pont-à-Mousson 1994
C Bandle, J Bemelmans, M Chipot, J Saint Jean Paulin and I Shafrir
327 Conjugate gradient type methods for ill-posed problems
M Hanke
328 A survey of preconditioned iterative methods
A M Bruaset
329 A generalized Taylor's formula for functions of several variables and certain of its applications
J-A Riestra
330 Semigroups of operators and spectral theory
S Kantorovitz
331 Boundary-field equation methods for a class of nonlinear problems
G N Gatica and G C Hsiao
332 Metrizable barrelled spaces
J C Ferrando, M López Pellicer and L M Sánchez Ruiz
333 Real and complex singularities
W L Marar
334 Hyperbolic sets, shadowing and persistence for noninvertible mappings in Banach spaces
B Lani-Wayda
335 Nonlinear dynamics and pattern formation in the natural environment
A Doelman and A van Harten
336 Developments in nonstandard mathematics
N J Cutland, V Neves, F Oliveira and J Sousa-Pinto
337 Topological circle planes and topological quadrangles
A E Schroth

Andreas E Schroth

TU Braunschweig, Germany

Topological circle planes and topological quadrangles

LONGMAN

Copublished in the United States with
John Wiley & Sons Inc., New York.

Longman Group Limited
Longman House, Burnt Mill, Harlow
Essex CM20 2JE, England
and Associated Companies throughout the world.

Copublished in the United States with
John Wiley & Sons Inc., 605 Third Avenue, New York, NY 10158

First published 1995

AMS Subject Classifications: (Main) 51H15, 51B05, 51E12
(Subsidiary) 51A50

ISSN 0269-3674

ISBN 0 582 28811 8

British Library Cataloguing in Publication Data

A catalogue record for this book is
available from the British Library

Library of Congress Cataloging-in-Publication Data

Schroth, A. (Andreas)
Topological circle planes and topological quadrangles / A. Schroth.
p. cm. -- (Pitman research notes in mathematics series, ISSN 0269-3674 ; ???)
1. Finite generalized quadrangles. 2. Circle. I. Title.
II. Series.
QA167.2.S37 1995
516'.13--dc20 95-22088
CIP

Printed and bound in Great Britain
by Biddles Ltd, Guildford and King's Lynn

Contents

Introduction

In this book the very close relationship between finite-dimensional connected circle planes and finite-dimensional connected generalised quadrangles is investigated. This relationship provides a better understanding of several problems in finite-dimensional connected circle planes. So, we have here an example of the phenomenon that new structures lead to a better understanding of older, seemingly well understood structures.

Here, the new structure is the concept of generalised quadrangles. Generalised quadrangles, or rather generalised polygons, were introduced by J. Tits in 1959 ([54]). Later they were subsumed under the even more general notion of "buildings" introduced in [53] to obtain a better understanding of certain groups.

The older concepts are, in this book, not groups but circle planes. The investigation of circles in the plane is almost as old as geometry itself. One of the most renowned ancient mathematicians who worked on circle planes is perhaps Apollonius of Perga (262 – 190 B.C.). He posed and solved the problem to construct a circle touching three given objects, each of which might be a line, a circle or a point. This "problem of Apollonius" fits nicely into the concept of generalised quadrangles and we will use this to successfully treat a topological version of the problem. Also, the original definition of a Laguerre plane in terms of oriented lines and oriented circles ([23]) can be described rather naturally in the setting of generalised quadrangles. Moreover, this description can be generalised and yields new connections between Laguerre planes and Möbius- or Minkowski planes.

Substantial results in geometry can usually only be obtained if some restrictions limit the geometries under investigation. So one studies e.g. geometries over fields or finite geometries or, as we do here, topological geometries. In all of these three cases there is a connection between circle planes and generalised quadrangles.

All that is needed to work out the connection between circle planes and generalised quadrangles over a field can be found in [10], even if the term generalised quadrangle is never used in this paper. And indeed, for the classical circle planes over the real or complex numbers the connection to quadrangles was neglected—although sometimes people came very close. It is well known that points and (oriented) circles of a circle plane are best treated as the points of a certain quadric ([33]). This quadric is the point space of the classical quadrangle. It was even noted that there is a natural relation on this quadric induced by the circle plane ([25]). But the fact that behind this relation hides a point-line geometry was missed. Perhaps in [34] Rigby comes closest to the concept of a point-line geometry. His paper is a convincing pleading that some problems in circle planes are best treated in the corresponding generalised quadrangle, only the term generalised quadrangle is never used.

The connection between finite circle planes and finite generalised quadrangles was investigated by Payne and Thas. In [30] they exhibit the connection between finite Laguerre planes and generalised quadrangles. In [51] Thas shows that there is a natural connection between finite Möbius- and Minkowski planes of even order and certain finite generalised quadrangles. However, to my knowledge, the connection between finite Möbius- and Minkowski planes of odd order and finite generalised quadrangles has not been studied successfully yet. The obstacle is that there seems to be no natural way to introduce oriented circles.

In this book we restrict our attention mainly to locally compact connected geometries. These are geometries where the sets of geometrical objects carry topologies such that they are locally compact connected spaces and such that all geometric operations are continuous. Since the real and complex numbers are locally compact fields the geometries over $\mathbb{R}$ and $\mathbb{C}$ are locally compact connected geometries. Hence, our investigations also include the classical cases.

To locally compact connected geometries we can usually assign topological parameters. Locally compact geometries with parameter 1 are, topologically, related to geometries over $\mathbb{R}$ while locally compact geometries with parameter 2 are related to geometries over $\mathbb{C}$. The main results of this book state that the connection between circle geometries with topological parameter s and generalised quadrangles with parameter s is as close as possible. More precisely, every locally compact circle plane with parameter s is related to a compact generalised quadrangle with parameter s, where $s \in \{1,2\}$. Conversely, for $s \in \{1,2\}$ every generalised quadrangle with parameter s is related to a Laguerre plane with parameter s. If, in addition, there is an involution on the quadrangle satisfying certain conditions, then the quadrangle is also related to a Möbius plane or to a Minkowski plane. It depends on the properties of the involution whether it is related to a Möbius plane or to a Minkowski plane. The involution corresponds to reversing orientation of oriented circles. The relationship between locally compact circle planes and compact generalised quadrangles is rather natural. One can introduce categories in an obvious way such that the constructions correspond to covariant functors.

The result that for $s \in \{1,2\}$ every generalised quadrangle with parameter s is related to a Laguerre plane with parameter s can also be formulated as follows: Up to duality every compact generalised quadrangle with parameter s, where $s \in \{1, 2\}$, is antiregular. Antiregularity is a purely geometric property. A generalised quadrangle is antiregular if any three pairwise non-collinear points are collinear with either precisely two points or no point. For compact generalised quadrangles this geometric property is closely related to topological properties. Namely, a compact generalised quadrangle with parameter s is antiregular if and only if there is a point such that the space of points collinear with but different from this point is homeomorphic to the point space of the classical locally compact Laguerre plane with parameter s.

To show that, up to duality, all compact generalised quadrangles with parameter s are antiregular two tools are introduced. The first tool is the notion of the twisting number of a compact generalised quadrangle. This number encodes the topological properties of the space of points collinear with a point. The second tool is the concept of square projections. In a compact generalised quadrangle with parameter s square projections are, topologically, maps from $\mathbb{S}_s$ to $\mathbb{S}_s$. The two tools twisting number and square projections are linked by the fact that the mapping degree of a square projection equals the twisting number.

Antiregularity of generalised quadrangles is closely related to the problem of Apollonius. This fact is used to completely solve the problem of Apollonius in locally compact connected circle planes.

Looking back now, the results presented here seem to be almost obvious. But the road to these results was neither straight nor easy. The first steps where made by S.E. Payne and J.A. Thas in [30]. They showed that all finite Laguerre planes of odd order are related to finite antiregular quadrangles. Inspired by this result M. Forst, who introduced topological generalised quadrangles, constructed compact quadrangles with parameter 1 from locally compact Laguerre- and Möbius planes with parameter 1.

In my Diplomarbeit ([36]) I tried to pursue the investigation of topological generalised quadrangles. I utterly failed to produce a compact quadrangle with parameter 1 not connected to a Laguerre plane with parameter 1. Instead, I showed that, up to duality, all compact quadrangles with parameter 1 are antiregular and hence connected to Laguerre planes ([36] or [37]). Already there I noted the connection between antiregularity and the problem of Apollonius ([38]). In fact, a paper by Groh, where the problem of Apollonius in Laguerre planes with parameter 1 is partly solved ([14]), inspired my attempt to prove that all locally compact quadrangles with parameter 1 are related to Laguerre planes. The proof as presented in [36] and [37] relies heavily on connectedness arguments. Thus hopes were low that the result could be extended to the higher parameter cases. The first attempt was only partly successful. In my Ph.D. thesis ([39]) I showed that locally compact Laguerre planes with parameter 2 lead to compact antiregular quadrangles with parameter 2 and that the quadrangles thus obtained are antiregular. Again, antiregularity of these quadrangles was linked to the solution of the problem of Apollonius in locally compact Laguerre planes with parameter 2.

However, whether or not all compact quadrangles with parameter 2 are connected to Laguerre planes still remained an open question. But the proofs in [39] already featured a certain map from traces to lines. A closer investigation of this map, now called square projection, provided the decisive clues for a complete solution. Also, results of L. Kramer published in [24] paved some of the way.

It was already noted by Forst in [11] that also locally compact Möbius planes with parameter 1 give rise to compact generalised quadrangles. However, to prove this he used a result by Groh ([15] and [16]). This result states that every locally compact Möbius plane with parameter 1 gives rise to a locally compact Laguerre plane with parameter 1. To obtain this result, Groh partly solved the problem of Apollonius in locally compact Möbius planes with parameter 1 ([17]). Interpreted in the setting of generalised quadrangles he proved that the quadrangle constructed from a Möbius plane with parameter 1 is 'locally' antiregular. But we know already that the quadrangle constructed from a Möbius plane is antiregular, so the result employed by Forst is stronger than necessary. The direct proof of the fact that one can construct generalised quadrangles from locally compact Möbius planes with parameter 1 offered no surprises. But it provided clues for the uncovering of the connection between locally compact connected Minkowski planes and generalised quadrangles. This connection completed the picture.

The book is organised as follows. Chapter 1 begins with the introduction of (topological) circle planes. The different models for the classical circle planes are discussed to some extent, since they provide ample motivation for the following constructions. The derived projective plane is also introduced. Some properties of the derived projective planes vital for the constructions developed in the following chapters are stated.

Chapter 2 is dedicated to generalised quadrangles and antiregular quadrangles in particular. Generalised quadrangles and square projections are introduced. Also a new invariant for compact quadrangles with parameter s, the twisting number, is developed. The interplay between this invariant and properties of square projections is used to characterise compact antiregular quadrangles. The main result states that for every compact antiregular quadrangle with parameters (s, t) we obtain $s = t \in \{1, 2\}$. Conversely, up to duality, every compact quadrangle with parameter 1 or 2 is antiregular.

The subject of the next three chapters is the connection between topological circle planes and generalised quadrangles. In Chapter 3 results on the relationship between finite-dimensional connected Laguerre planes and generalised quadrangles are presented. The main result of the previous chapter is used to give a new proof of the fact that the generalised quadrangle constructed from a locally compact Laguerre plane with parameter 2 is antiregular.

In Chapter 4 the connection between flat Möbius planes, i.e., a locally compact Möbius plane with parameter 1, and generalised quadrangles is investigated. It is shown that every flat Möbius plane gives rise to a compact quadrangle with parameter 1 featuring an involution with certain properties. Conversely, every such quadrangle can be obtained from a flat Möbius plane.

In Chapter 5 the connection between finite-dimensional connected Minkowski planes and generalised quadrangles is discussed. The main result concerning locally compact Minkowski planes with parameter 2 states that every such Minkowski plane gives rise to a compact quadrangle with parameter 2 featuring an involution with certain properties. Conversely, every such quadrangle can be linked to a locally compact Minkowski plane with parameter 2. The result for locally compact Minkowski plane with parameter 1 offers a slight surprise. It states that every such Minkowski plane gives rise to a pair of compact quadrangles with parameter 1, a pair of involutions, and a map between the two quadrangles. Again, the construction can be reversed.

The subject of Chapter 6 is the relationship between circle planes. First, the connections between circle planes and generalised quadrangles obtained so far are looked at from different points of view. From one point we see that the results of the previous chapters provide a method to construct a circle plane from another circle plane via a generalised quadrangle. The first section introduces the sister of a Laguerre plane, i.e., a method to construct a Laguerre plane from another Laguerre plane. The second section features a construction of a flat Laguerre plane from a flat Möbius plane. This construction generalises the original definition of a Laguerre plane via oriented circles. The construction appeared already in [15] and [16], but the present context provides a better understanding of the construction and also shorter proofs. In the final section of Chapter 6 a construction of a Laguerre plane from a Minkowski plane is discussed.

Chapter 7 is devoted entirely to the problem of Apollonius in finite-dimensional connected circle planes. The problem is posed and solved completely. The proofs use the fact that the quadrangles constructed from circle planes are antiregular. Moreover, square projections are employed extensively.

At first glance this book seems to be the end of the story. Well, it is not. There are still open questions, some of them are even prompted by the results presented here. The connection between finite Möbius- and Minkowski planes of odd order and generalised quadrangles remains open. Another rather irritatingly open problem is the following: Is the point space of a compact generalised quadrangle with parameter s always homeomorphic to the point- or line space of the corresponding classical generalised quadrangle? There is some hope that this problem can be solved using the twisting number and square projections. Another field of interest is the investigation of compact quadrangles with large groups of automorphisms. The results presented here, namely the connection between compact generalised quadrangles with parameter s and locally compact Laguerre planes with parameter s enable us to employ results on locally compact Laguerre planes with large groups of automorphisms. Here work is in progress.

The text is based on my habilitation thesis ([46]). To make the treatment more self-contained some material that appeared already elsewhere has been added. Also the classical real case is presented in more detail. I am very grateful for the opportunity to present the text as a whole. This way, the different constructions can be compared rather easily.

A work like this is hardly ever accomplished without the help of other persons. I would like to thank all those who have contributed in one way or other to this book. Special thanks go to Prof. R. Löwen and Prof H. Salzmann who introduced me to the world of topological geometry and saw to it that I didn't get lost in it. I am also thankful to Prof. J.Ch. Fisher who reported some omissions in the first version of this text. The text was prepared using programs by D.E. Knuth, namely TEX for typesetting and METAFONT to draw the figures.

1 Circle planes

Introduction

In this chapter we introduce the notion of (topological) circle planes and fix notation. Circle planes are also named Benz planes after W. Benz who first studied circle planes as a whole systematically. His approach, as for example exposed in [2], is mainly algebraic whereas the approach exposed here is more topological. Another very elegant unifying approach to circle planes was developed by F. Buekenhout in [6].

After the introduction of circle planes in general and the three different types of circle planes, namely Möbius, Laguerre and Minkowski planes, we discuss the derived or internal structure of a circle plane at some point s. We will see that this derived structure is an indispensable tool when dealing with circle planes. The crucial point is that circles not incident with s become ovals in the derived structure. In case of topological circle planes these circles become even topological ovals. This enables us to use the results on topological ovals obtained in [4].

1.1 Definitions and Notation

A **circle plane** $\mathfrak{Cp} = (\mathcal{S}, \mathcal{K}, \mathcal{E})$ is an incidence geometry with **point set** $\mathcal{S}$, **set of circles** $\mathcal{K}$ and a collection $\mathcal{E}$ of equivalence relations on $\mathcal{S}$ called **parallelisms** such that the following axioms hold, where two points are said to be **non-parallel** if they are not in relation for any of the equivalence relations in $\mathcal{E}$ and where two circles k and l are said to **touch**, if either $k = l$ or there is precisely one point incident with k and l:

C1: For any three pairwise non-parallel points $s, t, u \in \mathcal{S}$ there is precisely one circle $\boldsymbol{\alpha(s,t,u)} \in \mathcal{K}$ incident with these points.

C2: For any circle k, any point s incident with k and any point t not parallel to s, there is precisely one circle $\boldsymbol{\beta(s,k,t)}$ incident with s and t that touches k.

C3: For any point s, any circle k and any parallelism $\|_i \in \mathcal{E}$ different from the identity relation, there is a unique point $\boldsymbol{\gamma_i(s,k)}$ on k which is i-parallel to s.

C4: For any two non-parallel points s and t and for any two distinct parallelisms $\|_i, \|_j \in \mathcal{E}$ different from the identity relation, there is precisely one point $\boldsymbol{\delta_{i,j}(s,t)}$ which is i-parallel to s and j-parallel to t.

C5: Every circle is incident with at least three points and not all points in $\mathcal{S}$ are incident with one circle.

Notation: Usually we will consider circles as subsets of $\mathcal{S}$. For any point $s \in \mathcal{S}$ let $\boldsymbol{\Gamma(s)} := \{t \in \mathcal{S} \mid \exists \|_i \in \mathcal{E} : s \|_i t\}$ denote the set of points which are parallel to s for some parallelism in $\mathcal{E}$. The set $\{\Gamma(s) \mid s \in \mathcal{S}\}$ is denoted $\boldsymbol{\mathcal{G}}$. The identity relation is also called the **trivial parallelism**. A **flag** is an incident point-circle pair and the set of all flags is denoted $\boldsymbol{\mathcal{F}}$. If (s,k) is a flag, then the set of circles touching k in s is called the **touching pencil** spanned by s and k and denoted $\boldsymbol{B(s,k)}$. All circles of a touching pencil b have one point in common. We will call this point the **carrier** or **support** of b and denote it $\boldsymbol{\sigma(b)}$. A touching pencil b together with its support $\sigma(b)$ is called an **extended touching pencil** and denoted $\boldsymbol{\overline{B}(s,k)}$. The set of all touching pencils and the set of all extended touching pencils is denoted $\boldsymbol{\mathcal{B}}$ and $\boldsymbol{\overline{\mathcal{B}}}$ respectively. We also say that two circles are **tangent** if they touch. The set of circles incident with a point s is denoted $\boldsymbol{\mathcal{K}_s}$ and the set of circles incident with two points s and t is denoted $\boldsymbol{\mathcal{K}_{s,t}}$.

Usually we will only specify the point set and the set of circles of a circle plane. Parallel classes can be reconstructed as maximal sets of points, where no two points are contained in a common circle. Only in special situations we will specify the parallel classes or even the flag space.

A **topological circle plane** is a circle plane where the point set $\mathcal{S}$ and the set of circles $\mathcal{K}$ carry Hausdorff topologies such that the maps α, β, γ_i, and $\delta_{i,j}$ defined in the above axioms as well as the intersection of two circles are continuous operations. A topological circle plane is called **compact**, **locally compact** or **connected**, if the point space has the respective topological property. The circle space need not have the same topological properties, as will be seen in the subsequent discussion. A **finite-dimensional circle plane** is a locally compact topological circle plane whose point space is finite-dimensional. It should be stressed that by 'dimension' we always understand the topological dimension rather than the geometrical dimension. Two-dimensional circle planes are also called **flat** circle planes. The classical circle planes over the real or complex numbers are locally compact circle planes of dimension two or four respectively.

A **Minkowski plane** is a circle plane where $\mathcal{E}$ consists of two non-trivial parallelisms denoted $||_+$ and $||_-$ which are called plus-parallelism and minus-parallelism respectively. The plus- and minus-parallel class of a point s is denoted $\mathbf{\Gamma_+(s)}$ and $\mathbf{\Gamma_-(s)}$ respectively. So $\Gamma(s) = \Gamma_+(s) \cup \Gamma_-(s)$ denotes the set of points which are plus- or minus-parallel to the point s. The set of all plus-parallel classes and the set of all minus-parallel classes is denoted $\mathcal{G}_+$ and $\mathcal{G}_-$ respectively. Note that $\mathcal{G} \neq \mathcal{G}_+ \cup \mathcal{G}_-$. In Chapter 5 we will see that $\mathcal{G}$ is naturally homeomorphic to $\mathcal{G}_+ \times \mathcal{G}_-$ and thus to $\mathcal{S}$. The maps introduced in Axiom C3 are denoted γ_+ and γ_-. The map $\delta_{+,-}$ is abbreviated $\boldsymbol{\delta}$. Note that $\delta_{-,+}(s,t) = \delta_{+,-}(t,s) = \delta(t,s)$.

A **Laguerre plane** is a circle plane where $\mathcal{E}$ consists of one non-trivial parallelism, denoted $||$. Therefore $\Gamma(s)$ denotes just the parallel class of a point s and $\mathcal{G}$ denotes the set of all parallel classes. The map introduced in Axiom C3 is abbreviated γ. Axiom C4 does not apply for Laguerre planes, since there are no two different equivalence relations in $\mathcal{E}$.

A **Möbius plane** is a circle plane where $\mathcal{E}$ consists only of the identity relation. Therefore $\Gamma(s) = \{s\}$ and neither Axiom C3 nor Axiom C4 apply to Möbius planes.

We will see later, why these are the only circle planes, i.e., why there are no circle planes where $\mathcal{E}$ contains more than two equivalence relations. In the case of Minkowski and Laguerre planes one can weaken Axiom C5. Instead of all circles we need only require that one circle is incident with at least four points. For Möbius planes the set $\mathcal{G}$ can be identified with the point set $\mathcal{S}$ in a natural way. Also, for Minkowski planes the map $\Gamma : \mathcal{S} \to \mathcal{G}$ yields a natural identification. However, for Laguerre planes there is no such identification. This seemingly unimportant fact is a source for difficulties when one wants to make the quadrangle obtained from a finite-dimensional Laguerre plane a topological quadrangle (see Chapter 3 or [41]).

Topological Möbius planes where introduced by R. Wölk in [55], topological Laguerre planes where introduced by H.J. Groh in [12] and [13] and topological Minkowski planes where introduced by A. Schenkel in [35]. The original definition of topological Möbius planes as introduced by R. Wölk requires in addition that the set of pairs of properly intersecting circles is open. However, results of R. Löwen in [26] state that at least for finite-dimensional Möbius planes this follows from the other topological requirements. For a further exposition of topological circle planes the reader is referred to [50].

1.2 Models for classical circle planes

We now discuss different models for the classical circle planes, i.e., for circle planes over the real or complex numbers. This discussion will introduce most of the central aspects of the theory developed in this book. Another introduction to these models may be found in a recent article by J.Ch. Fisher ([10]). He looks at these models and the connections in between the different models for one circle plane over an arbitrary field.

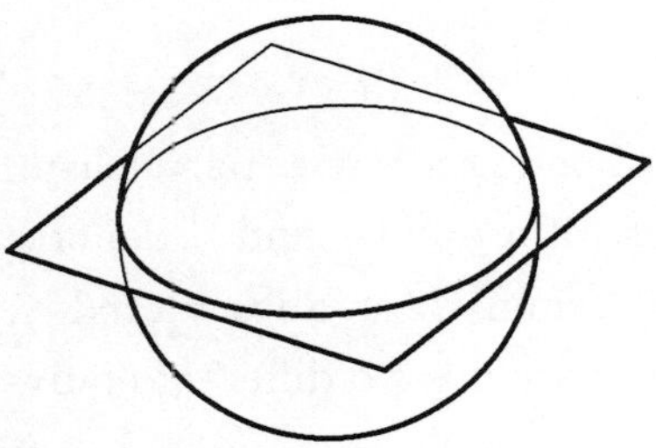

Möbius plane
Fig. 1.1.a

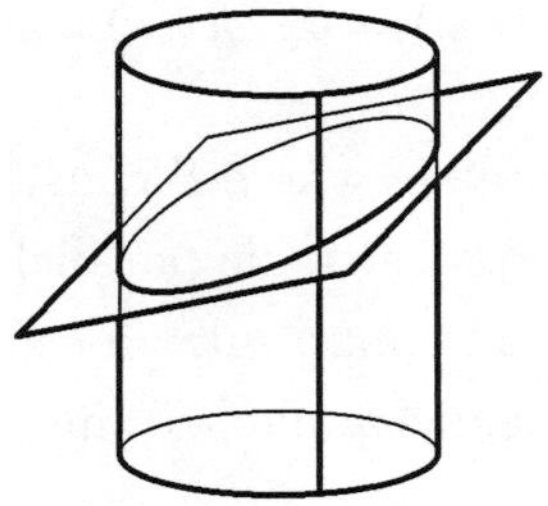

Laguerre plane
Fig. 1.1.b

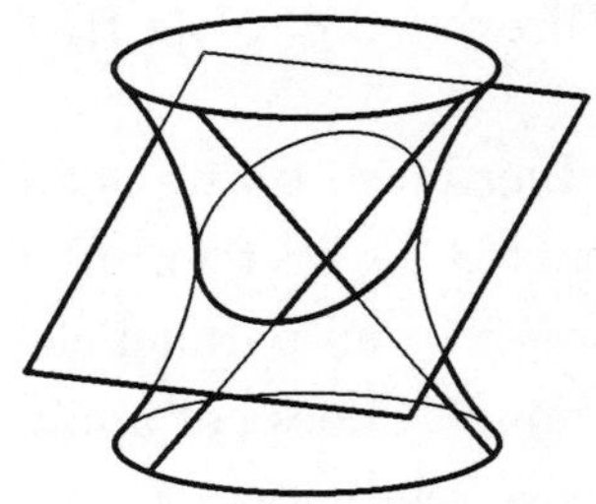

Minkowski plane
Fig. 1.1.c

One set of models for Möbius, Laguerre and Minkowski planes is obtained as the geometries of non-trivial plane sections of an elliptic quadric, an elliptic cone with its vertex removed, or a ruled quadric, respectively, in the three-dimensional projective space over some field. The so called classical models for finite-dimensional circle planes are obtained by taking the real or complex numbers in the above description. However, for Möbius planes there is no complex model, since there is no elliptic quadric in projective three-space over the complex numbers. We will see later, when we discuss the different types of circle planes, that topologically, finite-dimensional connected circle planes are very close to their classical counterparts.

Another set of models for the classical circle planes is obtained as follows. The Möbius plane is the geometry of lines and circles of the real Euclidean plane, the Laguerre plane is the geometry of lines and parabolas of the real or complex Euclidean plane and the Minkowski plane is the geometry of lines and hyperbolas of the real or complex Euclidean plane. The connection to the above models is given by projection.

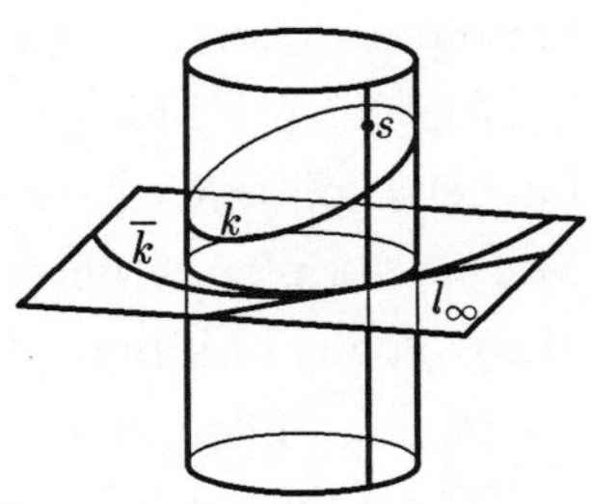

Fig. 1.2

More precisely, consider the point space as a quadric embedded in projective 3-space. Take a point s of the quadric and a subplane P of the projective space not containing the point s. Next project the circle geometry from the point s onto the subplane P. Let l_∞ denote the line of P obtained by intersecting P with the plane of tangency to the quadric at s. Let $A := P \setminus l_\infty$. All points parallel to s are projected to the line l_∞. All points not parallel to s are mapped bijectively to points of A. All circles through s become lines in A. In the case of Möbius planes, the projection of any circle not incident with s misses l_∞. In the case of Laguerre planes, the projections of all circles not incident with s intersect l_∞ in a common point. In the case of Minkowski planes, the projections of all circles not incident with s intersect l_∞ in two common points.

Choosing suitable coordinates for A, the projections of circles not incident with s become circles, parabolas or hyperbolas if the circle plane is a Möbius plane, a Laguerre plane or a Minkowski plane respectively. The topological version of this set of models will be explored in the next section.

As a link to quadrangles we now discuss a third way to set up models for the classical real circle planes. Consider the four-dimensional projective space $PG(4, \mathbb{R})$ over the real numbers equipped with the bilinear form φ associated to the matrix

$$\begin{pmatrix} 1 & & & & \\ & 1 & & & \\ & & 1 & & \\ & & & -1 & \\ & & & & -1 \end{pmatrix}$$

and let $Q := \{x \in PG(4, \mathbb{R}) \mid \varphi(x, x) = 0\}$ denote the quadric defined by φ. We use homogeneous coordinates in $PG(4, \mathbb{R})$. For $x = \langle(a, b, c, d, e)\rangle \in PG(4, \mathbb{R})$ let

$$H_x := \{y = \langle(y_1, y_2, y_3, y_4, y_5)\rangle \in PG(4, \mathbb{R}) \mid ay_1 + by_2 + cy_3 + dy_4 + ey_5 = 0\}$$

denote the hyperspace perpendicular to x with respect to the standard scalar product and let $x^* := \{y \in PG(4, \mathbb{R}) \mid \varphi(x, y) = 0\}$ denote the space of points perpendicular to x with respect to the form φ. Calculation shows that $H_p \cap Q$ is an elliptic quadric, if $\varphi(p, p) < 0$, that $H_p \cap Q$ is a singular quadric, if $\varphi(p, p) = 0$ and that $H_p \cap Q$ is a ruled quadric, if $\varphi(p, p) > 0$.

Since the group of collineations on $PG(4, \mathbb{R})$ respecting the form φ acts transitively on Q as well as on $\{x \mid \varphi(x, x) < 0\}$ and $\{x \mid \varphi(x, x) > 0\}$, one needs to check the above claim only for one suitable point per case. So, let $p_1 := \langle(0, 0, 0, 0, 1)\rangle$, $p_2 := \langle(1, 0, 0, 0, 1)\rangle$ and $p_3 := \langle(1, 0, 0, 0, 0)\rangle$. Then $\varphi(p_1, p_1) < 0 = \varphi(p_2, p_2) <$

$\varphi(p_3, p_3)$. Moreover:

$$\begin{aligned} H_{p_1} \cap Q &= \{\langle(a,b,c,d,0)\rangle \in PG(4,\mathbb{R}) \mid a^2+b^2+c^2=d^2\} \\ &= \{\langle(a,b,c,1,0)\rangle \in PG(4,\mathbb{R}) \mid a^2+b^2+c^2=1\}, \\ H_{p_2} \cap Q &= \{\langle(a,b,c,d,a)\rangle \in PG(4,\mathbb{R}) \mid b^2+c^2=d^2\} \\ &= \{p_2\} \cup \{\langle(a,b,c,1,a)\rangle \in PG(4,\mathbb{R}) \mid b^2+c^2=1\}, \\ H_{p_3} \cap Q &= \{\langle(0,a,b,c,d)\rangle \in PG(4,\mathbb{R}) \mid a^2+b^2=c^2+d^2\}. \end{aligned}$$

Thus $H_{p_1} \cap Q$ is isomorphic to the 2-sphere $\mathbb{S}_2$, the intersection $H_{p_2} \cap Q$ is a singular quadric where p_2 is the singular point and $H_{p_3} \cap Q$ is a ruled quadric.

Note that if $p \in Q$, i.e., if $\varphi(p,p)=0$, then $H_p \cap Q = p^* \cap Q$ and p is the singular point of the quadric $H_p \cap Q$. For $p \in Q$ we denote the quadric $H_p \cap Q$ by $p^\perp$.

Our next aim is to see the circle geometries within the quadric Q. If $\varphi(p,p)=0$ let $\Sigma_p := p^\perp \setminus \{p\}$ and if $\varphi(p,p) \neq 0$ let $\Sigma_p := H_p \cap Q$. Then

$$\begin{aligned} \mathfrak{Cp}(p) &:= (\Sigma_p, \{\Sigma_p \cap H_x \mid x \in Q \setminus H_p\}, \in) \\ &= (\Sigma_p, \{\Sigma_p \cap x^\perp \mid x \in Q \setminus p^\perp\}, \in) \end{aligned}$$

is almost a circle plane. More precisely, if $\varphi(p,p) < 0$ then $\mathfrak{Cp}(p)$ is a Möbius plane and if $\varphi(p,p) = 0$ then $\mathfrak{Cp}(p)$ is a Laguerre plane. If $\varphi(p,p) > 0$ then $\mathfrak{Cp}(p)$ has the point space of a Minkowski plane, but only half of the circles, namely only one of the two connected components of the circle space of a Minkowski plane. To obtain the missing circles one can take an involution ι on Σ_p respecting the form such that two non-intersecting lines contained entirely in Σ_p are fixed pointwise. Then

$$\begin{aligned} \overline{\mathfrak{Cp}}(p) &:= (\Sigma_p, \{\Sigma_p \cap H_x \mid x \in Q \setminus H_p\} \cup \{\iota(\Sigma_p \cap H_x) \mid x \in Q \setminus H_p\}, \in) \\ &= (\Sigma_p, \{\Sigma_p \cap x^\perp \mid x \in Q \setminus p^\perp\} \cup \{\iota(\Sigma_p \cap x^\perp) \mid x \in Q \setminus p^\perp\}, \in) \end{aligned}$$

is a Minkowski plane. A proof of these claims may be found in [10], or, preferably, the reader can verify them by calculations of his or her own. Again, it suffices to check the claims for the points p_1, p_2 and p_3 described above.

So far we have not used the fact that the quadric Q is a ruled quadric, i.e., that there are lines of the projective space entirely contained in Q. Let $\mathcal{L}$ denote the lines of $PG(4,\mathbb{R})$ contained entirely in Q. If x and y are two not necessarily distinct points in Q, then $\varphi(x,y)=0$ if and only if there is such a line containing x and y. Thus we obtain that $x^\perp$ consists precisely of those points in Q which are collinear with x. Hence, introducing lines allows us to neglect the bilinear form φ. This is the step from analytic geometry to incidence geometry.

Paraphrasing the above construction we obtain that for any $p \in Q$ the derived structure

$$\mathfrak{Cp}(Q,p) := (p^\perp \setminus \{p\}, \{p^\perp \cap q^\perp \mid q \in Q \setminus p^\perp\}, \in)$$

is a Laguerre plane. If τ is an involutory collineation on Q, i.e., an involution acting on Q and $\mathcal{L}$ that respects incidence, let Σ_τ denote the set of points fixed by τ. Depending on the form of Σ_τ we can construct a Möbius plane or a Minkowski plane. More precisely, if Σ_τ is an elliptic quadric, then

$$\mathfrak{Cp}(Q,\tau) := (\Sigma_\tau, \{\Sigma_\tau \cap q^\perp \mid q \in Q \setminus \Sigma_\tau\}, \in)$$

is a Möbius plane. If Σ_τ is a ruled quadric, and ι an involutory collineation on Σ_τ fixing two non-intersecting lines pointwise, then

$$\mathfrak{Cp}(Q,\tau) := (\Sigma_\tau, \{\Sigma_\tau \cap q^\perp \mid q \in Q \setminus \Sigma_\tau\} \cup \{\iota(\Sigma_\tau \cap q^\perp) \mid q \in Q \setminus \Sigma_\tau\}, \in)$$

is a Minkowski plane. As an example consider the involution τ_1 induced by the map $(a,b,c,d,e,f) \mapsto (a,b,c,d,e,-f)$ and the involution τ_2 induced by the map $(a,b,c,d,e,f) \mapsto (-a,b,c,d,e,f)$. Let p_1 and p_3 be as above. Then $\mathfrak{Cp}(Q,\tau_1) = \mathfrak{Cp}(p_1)$ and $\mathfrak{Cp}(Q,\tau_2) = \overline{\mathfrak{Cp}}(p_3)$, provided in the latter case the same involution ι is used.

The interesting point in these constructions in comparison to the constructions introduced earlier is that they work within the quadric Q. Moreover, the bilinear form is replaced by the concept of collinearity. The major task of this book is to extend this point of view. That is, we will work with geometric structures generalising the incidence structure of the above quadric, namely antiregular generalised quadrangles. We will see that at least in the finite-dimensional connected case, the above constructions work also and can even be reversed.

1.3 Derived structures

A very strong tool when dealing with circle planes is the derivation or internal structure of a circle plane at some point s. This is an incidence geometry whose points are the points of the circle plane $\mathfrak{Cp}$ not parallel to s and whose lines are all circles incident with s and all non-trivial parallel classes not containing s. It follows readily from the axioms of a circle plane, that this derived structure is an affine plane. More formally, if s is a point of a circle plane $\mathfrak{Cp} = (\mathcal{S}, \mathcal{K})$, let $\mathcal{G}'$ denote the set of all parallel classes containing at least two points but not s. Then the derived structure

$$\mathfrak{A}(\mathfrak{Cp}, s) := (\mathcal{S} \setminus \Gamma(s), \{k \in \mathcal{K} \mid s \in k\} \cup \mathcal{G}', \in)$$

is an affine plane, the **derived affine plane at s**. The projective closure of $\mathfrak{A}(\mathfrak{Cp}, s)$ is denoted $\mathfrak{P}(\mathfrak{Cp}, s)$ and called the **derived projective plane at s**. The line at infinity of $\mathfrak{P}(\mathfrak{Cp}, s)$, i.e., the additional line of $\mathfrak{P}(\mathfrak{Cp}, s)$, is denoted l_∞.

By construction, a circle l of $\mathfrak{Cp}$ incident with s becomes a line of the derived projective plane denoted $\bar{l}$. Before we determine what circles missing s become in the derived plane, we need the notion of (topological) ovals. An **oval O** in a projective plane $\mathfrak{P}$ is a set O of points, such that every line meets O in at most two points and such that through every point p in O there is a unique **tangent** to O at p, i.e., a unique line l such that $l \cap O = \{p\}$. Given an oval O in a projective plane, let $\mathcal{L}_O$ denote all lines intersecting O and let $O * O$ denote the symmetric product of O, i.e., the space obtained from $O \times O$ by identifying (x, y) with (y, x). The map $\psi_O : \mathcal{L}_O \to O * O : l \mapsto l \cap O$ is a bijection. A **topological oval** is an oval O where the map ψ_O is continuous with respect to the induced topologies.

Topological ovals in topological projective planes were introduced and discussed in detail in [4]. From there we cite the following result:

(1.1) Theorem: *Suppose O is a topological oval in a finite-dimensional connected projective plane $\mathfrak{P} = (\mathcal{P}, \mathcal{L})$. Then $\mathfrak{P}$ is either two- or four-dimensional.*

Moreover, if $\dim \mathcal{P} = 2$, then O is homeomorphic to the circle $\mathbb{S}_1$ and contained in an affine subplane $\mathfrak{A}$ of $\mathfrak{P}$. The point space A of the affine subplane is homeomorphic to $\mathbb{R}^2$ and the bounded connected component B of $A \setminus O$ consists precisely of those points which are not on a line that is tangent to O or that passes O. Every point on $\mathcal{P} \setminus \{B \cup O\}$ is incident with lines passing O, with lines intersecting O in two points and with exactly two tangents to O.

If $\dim \mathcal{P} = 4$, then O is homeomorphic to the sphere $\mathbb{S}_2$, there are no lines passing O and through every point not incident with O there are precisely two lines tangent to O.

By construction every non-trivial parallelism $\|_i \in \mathcal{E}$ of the circle plane determines a parallel class in the derived affine plane $\mathfrak{A}(\mathfrak{Cp}, s)$ and thus in the derived projective plane $\mathfrak{P}(\mathfrak{Cp}, s)$ every non-trivial parallelism $\|_i \in \mathcal{E}$ determines a unique point ∞_i on the line at infinity. Let $\Omega \subset l_\infty$ denote all points on the line at infinity corresponding to a non-trivial parallelism of the circle plane. If k is a circle not incident with the point s, let $\overline{k} := (k \setminus \Gamma(s)) \cup \Omega$. Any line in the derived projective plane, i.e., any circle through s, meets k in at most two points. For any point $t \in k$ the line $\overline{\beta(s,t,k)}$ is the unique tangent to k in t. Therefore, for every circle not incident with s the set $\overline{k}$ is an oval, provided Ω contains at most two points, i.e., provided there are at most two non-trivial parallelisms in $\mathcal{E}$.

For Möbius planes the set Ω is empty, so $l_\infty \cap \overline{k}$ is empty for every circle not incident with s and, as sets of points, $k = \overline{k}$. For Laguerre planes the set Ω consists of one point, thus l_∞ is tangent to $\overline{k}$ for every circle k missing s. For Minkowski planes the set Ω consists of two points, hence l_∞ intersects $\overline{k}$ in two points for every circle k not incident with s.

The connection between the number of non-trivial parallel relations and the number of points on $\overline{k}$ incident with l_∞ indicates why there are no circle planes with more than two non-trivial parallelisms. Note that a parabola in an affine plane has the line at infinity as a tangent line and that a hyperbola in an affine plane intersects the line at infinity in two points. So we generalised the models of circle planes as the geometries of circles, parabolas or hyperbolas in the plane.

In Chapter 5 we will have a closer look on the derived projective plane of a flat Möbius plane and in Chapter 6 we will inspect the derived projective plane of a finite-dimensional Minkowski plane in detail.

If $\mathfrak{Cp}$ is a finite-dimensional circle plane, then $\mathfrak{P}(\mathfrak{Cp}, s)$ is a topological projective plane and for every circle k not incident with s the induced oval $\overline{k}$ is closed, hence a topological oval ([49] and [4]: 2.6). Therefore, Theorem 1.1 implies the following result (cf. [4]: 3.2).

(1.2) Corollary: *Every finite-dimensional connected circle plane $\mathfrak{Cp} = (\mathcal{S}, \mathcal{K})$ is either two- or four-dimensional and there is no four-dimensional Möbius plane. Circles of a $2n$-dimensional circle plane are homeomorphic to the n-sphere $\mathbb{S}_n$.* ⟡

Notation: We say that a connected circle plane of dimension $2n$ has **topological parameter** n. The topological parameter of a circle plane is analogous to the order or parameter of a finite circle plane.

2 Antiregular quadrangles

Introduction

In this chapter we recall the definition of a (topological) generalised quadrangle. We also give a definition of antiregular quadrangles and state some results on antiregular quadrangles. A new invariant for compact quadrangles with topological parameter s, the twisting number, is established. Also a certain map, called square projection, is introduced. This map is quite natural for generalised quadrangles and it has particularly nice properties in regular or antiregular quadrangles. Moreover, it is closely related to the twisting number. Conversely, one can characterise antiregular quadrangles by properties of square projections. We will use this characterisation to show that up to duality all compact quadrangles with topological parameter 1 or 2 are antiregular.

Square projections will appear again in Chapter 7. There they will be a handy tool to solve the problem of Apollonius in topological circle planes. The twisting number can also be defined for compact n-gons with topological parameter s, but this is of no concern here.

2.1 Generalised quadrangles

A **generalised quadrangle** is an incidence structure $\mathfrak{GQ} = (\mathcal{P}, \mathcal{L}, \mathcal{F})$ of 'points', 'lines' and 'flags', i.e., incident point-line pairs, satisfying the following axioms:

Q1: Any two distinct points $p, q \in \mathcal{P}$ have at most one joining line, denoted $\boldsymbol{p \vee q}$.

Q2: For every anti-flag $(p, k) \in \mathcal{P} \times \mathcal{L} \setminus \mathcal{F}$ there exists exactly one flag $(q, l) \in \mathcal{F}$ such that $(p, l) \in \mathcal{F}$ and $(q, k) \in \mathcal{F}$.

Q3: Every point is incident with at least two lines, and dually.

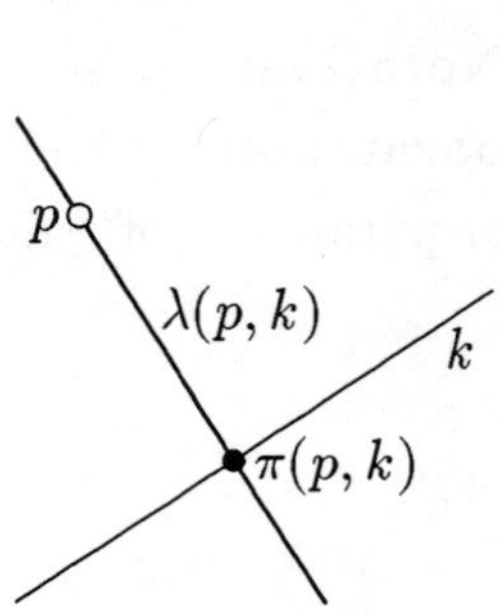

Fig. 2.1

Axiom Q2 yields two mappings $\boldsymbol{\pi} : \mathcal{P} \times \mathcal{L} \setminus \mathcal{F} \to \mathcal{P}$ and $\boldsymbol{\lambda} : \mathcal{P} \times \mathcal{L} \setminus \mathcal{F} \to \mathcal{L}$ with $\pi(p, k) = q$ and $\lambda(p, k) = l$.

A **thick** generalised quadrangle is a generalised quadrangle where every point is incident with at least three lines and where every line is incident with at least three points.

It is obvious that the dual structure of a (thick) generalised quadrangle is a (thick) generalised quadrangle as well. Again, two points x, y are said to be **collinear**, denoted $\boldsymbol{x \sim y}$, if they can be joined by a line. The set of points collinear with a given point p is denoted $\boldsymbol{p^\perp}$. The set of lines incident with a point p is called the **line pencil** through p and denoted $\boldsymbol{\mathcal{L}_p}$. The set of points incident with a line l is called the **point row** of l and denoted $\boldsymbol{\mathcal{P}_l}$. If p and q are two non-collinear points, the intersection $p^\perp \cap q^\perp$ is called the **trace** of p and q. For three points p, q and r the intersection $p^\perp \cap q^\perp \cap r^\perp$ is called the **centre** of p, q and r. A generalised quadrangle where no centre is empty is called a **centric** quadrangle. A set of three pairwise non-collinear points is called a **triad**. The space of all triads is denoted $\boldsymbol{\mathcal{T}}$. Generalised quadrangles are also known as "polar spaces of rank two". For a further introductory treatment of generalised quadrangles the reader is referred to [52] or [31].

A **topological quadrangle** is a generalised quadrangle where $\mathcal{P}$ and $\mathcal{L}$ carry Hausdorff topologies such that the maps π and λ become continuous. For an introduction to topological quadrangles we refer to [11] and [18]. A topological quadrangle is called a **compact quadrangle** if $\mathcal{P}$ and $\mathcal{L}$ carry compact topologies.

In this book we will mainly consider compact quadrangles whose point rows and line pencils are topological manifolds of positive dimension. If $\mathfrak{GQ}$ is such a quadrangle, then there are parameters $s, t \in \mathbb{N}$ such that all point rows are homeomorphic to the s-dimensional sphere $\mathbb{S}_s$ and all line pencils to the t-dimensional sphere $\mathbb{S}_t$ ([22]). In this case we speak of a compact quadrangle $\mathfrak{GQ}$ with **topological parameters $\boldsymbol{(s, t)}$** or a compact quadrangle $\mathfrak{GQ}$ of **order $\boldsymbol{(s, t)}$**. If $s = t$ we say the quadrangle $\mathfrak{GQ}$ has parameter s rather than parameters (s, s). Since we consider only compact quadrangles, we will usually drop the term 'topological'. At one stage we will also consider compact quadrangles where every point is incident with precisely two lines and every point row is homeomorphic to $\mathbb{S}_s$. We will call such a quadrangle a compact quadrangle with parameters $(s, 0)$. It is easy to verify that in these quadrangles the point space is homeomorphic to $\mathbb{S}_s \times \mathbb{S}_s$ and the line space is the disjoint union of two copies of $\mathbb{S}_s$. One might also define compact quadrangles with parameters $(0, t)$ and finally obtain that any compact quadrangle with parameters $(0, 0)$ is isomorphic to the common quadrangle, i.e., to the quadrangle consisting of four points and four lines. Note that the dual of a compact quadrangle with parameters (s, t) is a compact quadrangle with parameters (t, s). Compact quadrangles with parameter 1 are also known as three-dimensional quadrangles ([11]).

A reader familiar with finite generalised quadrangles will note that the definition of the order of a compact quadrangle differs from the definition of the order of a finite generalised quadrangle. For finite generalised quadrangles the order is simply the number of points on a line and the number of lines through a point ([31]). However, despite the difference in definition, the order of compact quadrangles behaves strikingly similar to the order of finite generalised quadrangles. An example for this can be found in [45].

There is also a rich theory of infinite totally disconnected compact quadrangles. The treatment of these quadrangles is completely different from the treatment of connected quadrangles and therefore well beyond the scope of this book.

We have already encountered a compact quadrangle. Namely the ruled quadric Q in real four-dimensional projective space discussed in the previous chapter as point set and all lines of the projective space which are contained in Q as line set forms a generalised quadrangle. This construction can be extended to arbitrary fields. Let $\mathbb{K}$ denote a field not of characteristic 2. Consider the symmetric form φ on $\mathbb{K}^5$ defined by the matrix

$$\begin{pmatrix} 1 & & & & \\ & 1 & & & \\ & & 1 & & \\ & & & -1 & \\ & & & & -1 \end{pmatrix}.$$

As points of the quadrangle take all 1-dimensional isotropic subspaces and as lines take all 2-dimensional totally isotropic subspaces of $\mathbb{K}^5$. This yields the **symmetric quadrangle $Q(4,\mathbb{K})$ over $\mathbb{K}$** ([31]). Note that two points of the quadrangle are collinear if and only if they are perpendicular with respect to φ. One verifies readily that lines as well as line pencils of the quadrangle $Q(4,\mathbb{K})$ are isomorphic to lines of the projective space $PG(4,\mathbb{K})$. Hence $Q(4,\mathbb{R})$ is a compact quadrangle with parameter 1 and $Q(4,\mathbb{C})$ is a compact quadrangle with parameter 2.

The dual of $Q(4,\mathbb{K})$ is the **symplectic quadrangle $W(\mathbb{K})$ over $\mathbb{K}$**. It can be constructed directly as follows: Consider the three-dimensional projective space $PG(3,\mathbb{K})$ equipped with a symplectic polarity. Points of the quadrangle are all points of $PG(3,\mathbb{K})$ and lines of the quadrangle are all totally isotropic lines. For a further discussion and characterisation of symplectic quadrangles see [45].

That the symplectic quadrangle $W(\mathbb{K})$ is indeed the dual to the symmetric quadrangle $Q(4,\mathbb{K})$ is for example proved in [31]. Although these two quadrangles are dual to each other, they are utterly different in nature. The symplectic quadrangle is the typical example of a 'regular quadrangle' (see [45]), whereas the symmetric quadrangle $Q(4,\mathbb{K})$ has the property that the centre of any triad is either empty or contains precisely two points. Generalised quadrangles with this property are called **antiregular** quadrangles. We will see in the next chapter that antiregular quadrangles are closely related to Laguerre planes.

The next result is, without reference, much used in this book. A proof can be found in [45]: 2.2.

(2.1) Lemma: *For any two non-collinear points p and q of a compact quadrangle with parameters (s,t) the trace $p^\perp \cap q^\perp$ is homeomorphic to $\mathcal{L}_p$, hence to $\mathbb{S}_t$.*

2.2 Square projections

In antiregular quadrangles the following restriction of the map π plays an essential rôle. Suppose p, q are two non-collinear points and suppose g is a line of a quadrangle such that no point of the trace $p^\perp \cap q^\perp$ is incident with g. Then the restriction

$$\mu_{p,q,g} : p^\perp \cap q^\perp \to g : x \mapsto \pi(x,g)$$

is called the **square projection** from $p^\perp \cap q^\perp$ to g. One immediately obtains

$$\mu^{-1}_{p,q,g}(\pi(p,g)) = \pi(q,\lambda(p,g)) \quad \text{and} \quad \mu^{-1}_{p,q,g}(\pi(q,g)) = \pi(p,\lambda(q,g)).$$

Apart from this, little can be said about square projections in general. Regularity or antiregularity of a quadrangle however imposes severe restrictions on the behaviour of square projections. In regular quadrangles square projections are bijections and in the topological case even homeomorphisms as can be deduced from the results in [45]. In antiregular quadrangles we see that for any point $z \in g \setminus \{\pi(p,g),\pi(q,g)\}$ the preimage $\mu^{-1}_{p,q,g}(z) = p^\perp \cap q^\perp \cap z^\perp$ contains either two or no points.

For compact quadrangles with parameter s we can obtain some information on square projections. If in addition the quadrangle is antiregular we can even determine square projections up to topological equivalence. Our main tool in these investigations is the concept of the degree of a continuous map from $\mathbb{S}_n$ to $\mathbb{S}_n$. The mapping degree can be defined as follows.

Let $f : \mathbb{S}_n \to \mathbb{S}_n$, where $n \geq 1$, denote a continuous map. Both, the nth homotopy group $\pi_n(\mathbb{S}_n)$ of $\mathbb{S}_n$ as well as the nth homology group $H_n(\mathbb{S}_n)$ of $\mathbb{S}_n$ are isomorphic to $\mathbb{Z}$. The map f induces a homomorphism $f_* : \pi_n(\mathbb{S}_n) = \mathbb{Z} \to \pi_n(\mathbb{S}_n) = \mathbb{Z}$ and

deg(*f*) $:= f_*(1) = m \in \mathbb{Z}$ is called the **mapping degree** of f. If the generators of the homotopy groups can be chosen independently, then the mapping degree is only defined up to sign.

This definition is of a rather global nature. Using homology groups rather than homotopy groups one sees that the mapping degree can also be determined locally. Let Σ denote a triangulation of $\mathbb{S}_n$ and let $s : \Sigma^t \to \Sigma$, where Σ^t is the tth barycentric subdivision of Σ, denote a simplicial approximation to f. Furthermore, let the simplices of Σ and Σ^t be oriented consistently. Let σ denote any simplex of Σ. Let T^+ denote the collection of simplices $\tau \in \Sigma^t$ with $s(\tau) = \sigma$ and let T^- denote the collection of simplices $\tau \in \Sigma^t$ with $s(\tau) = -\sigma$ where $-\sigma$ denotes the simplex σ with reversed orientation. Then $\deg(f) = |\mathrm{T}^+| - |\mathrm{T}^-|$ (see: [1]: Theorem 9.1).

The next lemma states some results on the mapping degree that we will use in the subsequent discussion.

(2.2) Lemma: *Let $f : \mathbb{S}_n \to \mathbb{S}_n$, where $n \geq 1$, denote a continuous map.*

a) If f is not surjective, then $\deg(f) = 0$.
b) If $n = 1$, then $|f^{-1}(z)| \geq \deg(f)$ for every $z \in \mathbb{S}_1$.
c) Suppose $n \geq 2$. If there are two points $x, y \in \mathbb{S}_n$ such that $|f^{-1}(x)| = |f^{-1}(y)| = 1$ and such that the restriction $f^\flat : \mathbb{S}_n \setminus \{f^{-1}(x), f^{-1}(y)\} \to \mathbb{S}_n \setminus \{x, y\}$ induces a homomorphism $f^\flat_* : \pi_{n-1}(\mathbb{S}_n \setminus \{f^{-1}(x), f^{-1}(y)\}) = \mathbb{Z} \to \pi_{n-1}(\mathbb{S}_n \setminus \{x, y\}) = \mathbb{Z}$ of the form $z \mapsto mz$, then $deg(f) = m$.
d) Suppose $n = 2$ and $\deg(f) > 1$. Then there are at most two points whose preimage consists of one point only. If there are two points $x, y \in \mathbb{S}_2$ such that $|f^{-1}(x)| = |f^{-1}(y)| = 1$, then $|f^{-1}(z)| \geq \deg(f)$ for every point $z \in \mathbb{S}_2 \setminus \{x, y\}$.

Proof: a) Suppose there is some $x \in \mathbb{S}_n \setminus f(\mathbb{S}_n)$. Then the map f can be factored as follows:

$$\mathbb{S}_n \xrightarrow{f} \mathbb{S}_n \setminus \{x\} \hookrightarrow \mathbb{S}_n.$$

On the level of homotopy groups this induces

$$\pi_n(\mathbb{S}_n) = \mathbb{Z} \xrightarrow{f_*} \pi_n(\mathbb{S}_n \setminus \{x\}) = 0 \hookrightarrow \pi_n(\mathbb{S}_n) = \mathbb{Z}$$

which implies the result.

b) and c) The results follows readily from the local characterisation of the mapping degree.

d) Suppose there are three points $x, y, z \in \mathbb{S}_2$ such that $|f^{-1}(x)| = |f^{-1}(y)| = 1$ and such that $|f^{-1}(z)| = k < m := \deg(f)$. Without loss of generality we may assume $f^{-1}(x) = x$ and $f^{-1}(y) = y$. Let $z_1, \ldots, z_k$ denote the elements of $f^{-1}(z)$. There are disjoint neighbourhoods U_x, U_y and U_i of x, y and z_i, respectively, homeomorphic to $\mathbb{R}^2$. There is a neighbourhood U_z of z homeomorphic to $\mathbb{R}^2$ and disjoint to U_x and U_y. Moreover, these neighbourhoods can be chosen such that there is a map $\varphi_x : \mathbb{S}_1 \to U_x \setminus \{x\}$, a map $\varphi_y : \mathbb{S}_1 \to U_y \setminus \{y\}$, a map $\varphi_z : \mathbb{S}_1 \to U_z \setminus \{z\}$ and maps $\varphi_i : \mathbb{S}_1 \to U_i \setminus \{z_i\}$ with the following properties: The map φ_x represents a generator of $\pi_1(U_x \setminus \{x\})$ and $f(\varphi_x(\mathbb{S}_1)) \subset U_x$, the map φ_y represents a generator of $\pi_1(U_y \setminus \{y\})$ and $f(\varphi_y(\mathbb{S}_1)) \subset U_y$, the map φ_z represents a generator of $\pi_1(U_z \setminus \{z\})$ and, for every $i \in \{1, \ldots, k\}$, the map φ_i represents a generator of $\pi_1(U_i \setminus \{z_i\})$ and $f(\varphi_i(\mathbb{S}_1)) \subset U_z$.

Since the degree of f is m we obtain by Part c:

$$\begin{aligned} [f \circ \varphi_x] &= m \in \pi_1(U_x \setminus \{x\}) \cong \mathbb{Z}, \\ [f \circ \varphi_y] &= m \in \pi_1(U_y \setminus \{y\}) \cong \mathbb{Z}, \\ \sum_{i=1}^{k} [f \circ \varphi_i] &= m \in \pi_1(U_z \setminus \{z\}) \cong \mathbb{Z}, \end{aligned}$$

where $[t]$ denotes the element of the fundamental group represented by t. Let l_i denote the element of $\pi_1(U_z \setminus \{z\})$ represented by $f \circ \varphi_i$. Then, by the above, $\sum_{i=1}^{k} l_i = m$. Since $k < m$ there is at most one $i \in \{1, \ldots, k\}$ such that $l_i > 1$.

Now consider the fundamental group $\pi_1(\mathbb{S}_2 \setminus \{x, y, z_1, \ldots, z_k\})$. This is the free group generated by $k + 1$ elements. As set of generators we may take the set $\{[\varphi_x], [\varphi_y], [\varphi_1], \ldots, [\varphi_{k-1}]\}$. We can choose the maps φ_i in such a way that we obtain $\prod_{i=1}^{k} [\varphi_i] = [\varphi_x][\varphi_y]$. (We use multiplication for the operation in a non-abelian group but addition for the operation in an abelian group.) The fundamental group $\pi_1(\mathbb{S}_2 \setminus \{x, y, z\})$ is the free group generated by two elements. As generators we take $[\varphi_x]$ and $[\varphi_y]$. Again, the map φ_z can be choosen such that $[\varphi_z] = [\varphi_x][\varphi_y]$.

The fundamental groups in these considerations are not commutative. Therefore we have to be careful about base points. We choose any $b \in \mathbb{S}_2 \setminus \{x, y, z_1, \ldots, z_k\}$ as base point for $\pi_1(\mathbb{S}_2 \setminus \{x, y, z_1, \ldots, z_k\})$ and $f(b)$ as base point for $\pi_1(\mathbb{S}_2 \setminus \{x, y, z\})$. We then have to 'connect' the maps φ_x, φ_y, φ_z and φ_i with paths to the base points. For φ_x and φ_y the connecting paths in $\mathbb{S}_2 \setminus \{x, y, z_1, \ldots, z_k\}$ might differ from the paths in $\mathbb{S}_2 \setminus \{x, y, z\}$. Since $[f \circ \varphi_x] = m \in \pi_1(U_x \setminus \{x\})$ and $[f \circ \varphi_y] = m \in \pi_1(U_y \setminus \{y\})$ we may choose the connecting paths such that $[f \circ \varphi_x] = f_*([\varphi_x]) =$

$[\varphi_x]^m$ and $[f \circ \varphi_y] = f_*([\varphi_y]) = [\varphi_y]^m$ where

$$f_* : \pi_1(\mathbb{S}_2 \setminus \{x, y, z_1, \ldots, z_k\}) \to \pi_1(\mathbb{S}_2 \setminus \{x, y, z\})$$

is the homomorphism induced by f. We now look at the image of $\prod_{i=1}^k [\varphi_i]$ under the homomorphism f_*. On the one hand we obtain

$$f_*(\prod_{i=1}^k [\varphi_i]) = \prod_{i=1}^k (f_*([\varphi_i])) = \prod_{i=1}^k (\alpha_i([\varphi_x][\varphi_y])^{l_i} \alpha_i^{-1}),$$

where α_i is some element of $\pi_1(\mathbb{S}_2 \setminus \{x, y, z\})$. On the other hand we have

$$f_*(\prod_{i=1}^k [\varphi_i]) = f_*([\varphi_x][\varphi_y]) = f_*([\varphi_x]) f_*([\varphi_y]) = [\varphi_x]^m [\varphi_y]^m.$$

Thus $[\varphi_x]^m [\varphi_y]^m = \prod_{i=1}^k (\alpha_i([\varphi_x][\varphi_y])^{l_i} \alpha_i^{-1})$. However, if $m > 2$ and $k < m$, then the condition $\sum_{i=1}^k l_i = m$ implies that any reduced word in $\pi_1(\mathbb{S}_2 \setminus \{x, y, z\})$ obtained from the word $\prod_{i=1}^k (\alpha_i([\varphi_x][\varphi_y])^{l_i} \alpha_i^{-1})$ contains a subword of the form $[\varphi_x]^{\varepsilon_1} [\varphi_y]^{\varepsilon_2} [\varphi_x]^{\varepsilon_3}$ or of the form $[\varphi_y]^{\varepsilon_1} [\varphi_x]^{\varepsilon_2} [\varphi_y]^{\varepsilon_3}$, where $\varepsilon_i \in \{1, -1\}$. Thus the equality $\prod_{i=1}^k (\alpha_i([\varphi_x][\varphi_y])^{l_i} \alpha_i^{-1}) = [\varphi_x]^m [\varphi_y]^m$ cannot hold. This contradiction completes the proof. ♋

Remark: Part d of the above lemma becomes false if $n > 2$. The proof of Part d becomes much easier if we consider only the case $\deg(f) = 2$. We will see later (Lemma 2.3 and Lemma 2.8) that this is the only case we are really interested in.

2.3 The twisting number

In this section we introduce a numerical invariant for compact quadrangles with parameter s. We call this invariant the twisting number of the quadrangle. The twisting number encodes a construction of the space $p^\perp \setminus (\{p\} \cup q^\perp)$ where p and q are two non-collinear points.

We begin our discussion with rather general considerations. First, we recall some results on the topological structure of the space $p^\perp \setminus \{p\}$. For a more detailed treatment of this topic the reader is referred to [40] or [22].

Suppose we are given a compact quadrangle with parameters (s, t). We will see below that for any point p the restriction

$$f : p^\perp \setminus \{p\} \to \mathcal{L}_p : x \mapsto x \vee p$$

is a locally trivial fibring. The base space is homeomorphic to the t-dimensional sphere. Fibres are punctured lines, hence fibres are homeomorphic to $\mathbb{R}^s$. If q is a point not collinear with p, then the map

$$f : p^\perp \setminus (\{p\} \cup q^\perp) \to \mathcal{L}_p : x \mapsto x \vee p$$

is still a fibration. This time however, fibres are homeomorphic to $\mathbb{R}^s \setminus \{0\}$, hence homotopy equivalent to the sphere of dimension $s-1$. This implies that the fibring f is homotopic to a fibring with base space $\mathbb{S}_t$ and fibres homeomorphic to $\mathbb{S}_{s-1}$. These fibrings are well understood ([47]: § 26).

Since for any two pairs (p,q) and (x,y) of points with $p \not\sim q$ and $x \not\sim y$ there exists a fibre preserving homeomorphism from $p^\perp \setminus (\{p\} \cup q^\perp) \to x^\perp \setminus (\{x\} \cup y^\perp)$ (see [40]) the above fibring is an invariant of the quadrangle.

In the present context the situation is reasonably nice, so that we do not have to use the full machinery of homotopy theory employed in [47]: § 26 in order to determine the possible fibrations. The idea is to obtain the bundle $x^\perp \setminus (\{x\} \cup y^\perp)$ via some gluing process. The gluing map eventually leads to the desired numerical invariant.

A **hemisphere** of a sphere $\mathbb{S}_n$ is a subspace H of $\mathbb{S}_n$ homeomorphic to the n-dimensional disc $\mathbb{D}_n$ such that the boundary $\partial H \cong \mathbb{S}_{n-1}$ is embedded tamely in $\mathbb{S}_n$, such that the closure of the complement $\overline{\mathbb{S}_n \setminus H}$ is also homeomorphic to $\mathbb{D}_n$ and such that $H \cap \overline{\mathbb{S}_n \setminus H} = \partial H$. Note that $\overline{\mathbb{S}_n \setminus H}$ is again a hemisphere, called the **hemisphere opposite** $\boldsymbol{H}$ and denoted $\boldsymbol{H^-}$.

Let p, q with $p \not\sim q$ be two points fixed throughout this section. Let $g_1, g_2 \in \mathcal{L}_p$ denote two lines through p. Let $x_i := \pi(q, g_i)$ and $h_i := \lambda(q, g_i)$; $i = 1, 2$.

The line pencil $\mathcal{L}_p$ is homeomorphic to the t-dimensional sphere $\mathbb{S}_t$. Let G_1 denote a hemisphere of $\mathcal{L}_p$ containing g_1 in its interior but not containing g_2. Let $G_2 := G_1^-$ denote the hemisphere opposite G_1. Let

$$X_i := \{x \in p^\perp \setminus \{p\} \mid x \vee p \in G_i\} = f^{-1}(G_i) \quad \text{for} \quad i \in \{1, 2\}$$

and let $Y := X_1 \cap X_2 = f^{-1}(G_1 \cap G_2)$.

For $i \in \{1, 2\}$ the space X_i is naturally homeomorphic to $h_{3-i} \setminus \{x_{3-i}\} \times G_i$ via the map

$$\kappa_i : h_{3-i} \setminus \{x_{3-i}\} \times G_i \to X_i : (x, g) \mapsto \pi(x, g)$$

whose inverse is

$$\kappa_i^{-1} : X_i \to h_{3-i} \setminus \{x_{3-i}\} \times G_i : y \mapsto (\pi(y, h_{3-i}), y \vee p).$$

If we take $\mathcal{L}_p \setminus \{g_{3-i}\}$ instead of G_i we obtain that the map f introduced above is indeed a locally trivial fibration.

To investigate this fibring further, choose a line $g \in G_1 \cap G_2$ and let

$$\xi : h_1 \to h_2 : x \mapsto \pi(\pi(x, g), h_2).$$

Then ξ is a homeomorphism that fixes q and that maps x_1 to x_2. Now consider the fibre preserving homeomorphism

$$\varphi : Y \to Y : x \mapsto \pi(\xi(\pi(x, h_1)), x \vee p)$$

whose inverse is

$$\varphi^{-1} : Y \to Y : x \mapsto \pi(\xi^{-1}(\pi(x, h_2)), x \vee p).$$

By construction, the map φ fixes $q^\perp \cap Y$ pointwise and therefore it fixes every fibre. In addition, φ fixes also $\mathcal{P}_g \setminus \{p\} \subset Y$ pointwise. We can reconstruct $p^\perp \setminus \{p\}$ once we know φ. Take two copies of $\mathbb{R}^s \times \mathbb{D}_t$ and glue them together along $\mathbb{R}^s \times \mathbb{S}_{t-1}$ using the map φ on $Y \cong \mathbb{R}^s \times \mathbb{S}_{t-1}$.

We do not need to know all of φ to reconstruct $p^\perp \setminus \{p\}$ topologically. It is enough to know how, up to homotopy, φ acts on $Y \setminus q^\perp$. The space $Y \setminus q^\perp$ is homeomorphic to $\mathbb{R}^s \setminus \{0\} \times \mathbb{S}_{t-1}$ which in turn is homotopy equivalent to $\mathbb{S}_{s-1} \times \mathbb{S}_{t-1}$.

In order to assign to the map φ a numerical invariant we have to leave these more general considerations and restrict ourselves to compact quadrangles with parameter s. In these cases the action of φ on $Y \setminus q^\perp$ can be classified quite easily.

First assume $s \geq 2$. Then the space $Y \setminus q^\perp$ is naturally homeomorphic to $\mathbb{R}^s \setminus \{0\} \times \mathbb{S}_{s-1}$. Thus $\pi_{s-1}(Y \setminus q^\perp) = \mathbb{Z} \times \mathbb{Z}$. Since φ fixes every fibre, the induced automorphism φ_* of $\mathbb{Z} \times \mathbb{Z}$ has the form $\varphi_*(a, b) = (na + mb, b)$. Since φ fixes one fibre pointwise, we may assume $\varphi_*(1, 0) = (1, 0)$. Therefore, the induced automorphism is of the form $\varphi_*(a, b) = (a + mb, b)$ where $m \in \mathbb{Z}$. After possibly interchanging the generator of one of the factors with its inverse, we may assume $m \in \mathbb{N}_0$.

Using Van Kampen's theorem if $s = 2$ (e.g.: [1]: 6.13) and the Mayer-Vietoris-sequence and Hurewicz theorem if $s > 2$ one can calculate the $s - 1^{\text{st}}$ homotopy group of $p^\perp \setminus (\{p\} \cup q^\perp)$ once one knows the automorphism φ_*. One obtains:

$$\pi_{s-1}(p^\perp \setminus (\{p\} \cup q^\perp)) = \mathbb{Z}/m\mathbb{Z}.$$

For $s = 2$ the precise argument runs as follows: For $i \in \{1,2\}$ let $X_i' = X_i \setminus q^\perp$. Let $\kappa_i : \mathbb{R}^2 \setminus \{0\} \times \mathbb{D}_2 \to X_i'$ denote the homeomorphisms induced by the coordinatisation. Let $Y_1' := \kappa_i(\mathbb{R}^2 \setminus \{0\} \times \mathbb{S}_1)$ and let $\iota_i : Y_i' \hookrightarrow X_i'$ denote the canonical inclusion. The fundamental group $\pi_1(X_i')$ is the infinite cyclic group $\mathbb{Z}$ generated by the element γ_i which is represented by the map $\rho_i : \mathbb{S}_1 \to X_i' : t \mapsto \kappa_i((0,1),t)$. The fundamental group $\pi_1(Y_i')$ is the group $\mathbb{Z}^2$ generated by the elements γ_i and δ_i where γ_i is as above and where δ_i is represented by the map $\psi_i : \mathbb{S}_1 \to Y_i' : t \mapsto \kappa_i(t,(0,1))$. Then, on the level of homotopy groups, we have: $\iota_{i*}(\gamma_i) = \gamma_i$ and $\iota_{i*}(\delta_i) = 0$.

We may identify $Y' := X_1' \cap X_2'$ with Y_1'. Then $Y' = \varphi(Y_2')$. By assumption we know $\varphi_*(a,b) = (a+mb, b)$ or, in terms of generators, $\varphi_*(\gamma_2) = \gamma_1$ and $\varphi_*(\delta_2) = m\gamma_1 + \delta_1$. For the inclusion $\iota_* = (\iota_2 \circ \varphi^{-1})_*$ of the generators of $\pi_1(Y') = \pi_1(Y_1')$ into $\pi_1(X_2')$ we therefore obtain:

$$\iota_*(\gamma_1) = \iota_{2*}(\varphi_*^{-1}(\gamma_1)) = \iota_{2*}(\gamma_2) = \gamma_2$$

and

$$\iota_*(\delta_1) = \iota_{2*}(\varphi_*^{-1}(\delta_1)) = \iota_{2*}(-m\gamma_2 + \delta_2) = -m\gamma_2.$$

By Van Kampen's theorem ([1]: 6.13) this implies that the fundamental group $\pi_1(p^\perp \setminus (\{p\} \cup q^\perp)) = \pi_1(X_1' \cup_\varphi X_2')$ is the group generated by γ_1 and γ_2 subject to the relations $\gamma_1 = \iota_{1*}(\gamma_1) = \iota_*(\gamma_1) = \gamma_2$ and $0 = \iota_{1*}(\delta_1) = \iota_*(\delta_1) = -m\gamma_2$. So, $\pi_1(p^\perp \setminus (\{p\} \cup q^\perp))$ is the group generated by one element γ subject to the relation $m\gamma = 0$.

Thus the number $m = m(\mathfrak{GQ})$ is a topologically quite sharp invariant of the quadrangle in question. More precisely, for two compact quadrangles with parameter s this number is the same if and only if for every point p of the one quadrangle and any point q of the other quadrangle, the spaces $p^\perp \setminus \{p\}$ and $q^\perp \setminus \{q\}$ are homeomorphic. We will call the number $m(\mathfrak{GQ})$ the **twisting number** of the quadrangle $\mathfrak{GQ}$. Since for the same parameter s but different twisting numbers m the spaces $p^\perp \setminus (\{p\} \cup q^\perp)$ are not homeomorphic and, since for any two pairs (p,q) and (x,y) of non-collinear points of a compact quadrangle the spaces $p^\perp \setminus (\{p\} \cup q^\perp)$ and $x^\perp \setminus (\{x\} \cup y^\perp)$ are homeomorphic, the number m does neither depend on the particular choice of p and q nor on the choice of the lines h_i used in the construction.

If the compact quadrangle has parameter 1, then $Y \setminus q^\perp$ consists of four connected components which we can identify in a natural way with $\mathbb{S}_0 \times \mathbb{S}_0$. The map φ induces a permutation φ_* on $\mathbb{S}_0 \times \mathbb{S}_0$ with $\varphi_*(\mathbb{S}_0 \times \{x\}) = \mathbb{S}_0 \times \{x\}$ for every $x \in \mathbb{S}_0$ and such that there is an element $y \in \mathbb{S}_0$ for which the restriction of φ_* to $\{y\} \times \mathbb{S}_0$ is the identity.

More explicitly, let $\mathbb{S}_0 = \{a, b\}$. Then we may assume $\varphi_*(a, a) = (a, a)$ and $\varphi_*(b, a) = (b, a)$. This leaves two possibilities. Either φ_* is the identity or φ_* interchanges (a, b) and (b, b). Both cases occur. The first case corresponds to generalised quadrangles where the space $p^\perp \setminus \{p\}$ is homeomorphic to the cylinder whereas the second case corresponds to generalised quadrangles where the space $p^\perp \setminus \{p\}$ is homeomorphic to the Möbius strip. In the first case we set $m(\mathfrak{G}\mathfrak{Q}) = 0$ and in the latter case we set $m(\mathfrak{G}\mathfrak{Q}) = 1$. It is shown in [39] that if p is a point of a compact quadrangle with parameter 1 and l a point of the dual quadrangle, then $p^\perp \setminus \{p\}$ is not homeomorphic to $l^\perp \setminus \{l\}$. Thus the twisting number of a compact quadrangle with parameter 1 is never equal to the twisting number of the dual quadrangle.

A recent result of Linus Kramer [24]: 4.3.1 states that for every compact quadrangle with parameter 2 the space $p^\perp \setminus \{p\}$ is either homeomorphic to the complex Hopf bundle or to the tangent bundle of $\mathbb{S}_2$. Moreover, the space $p^\perp \setminus \{p\}$ is homeomorphic to the complex Hopf bundle if and only if for any point l of the dual quadrangle the space $l^\perp \setminus \{l\}$ is homeomorphic to the tangent bundle of $\mathbb{S}_2$. If the space $p^\perp \setminus \{p\}$ is homeomorphic to the complex Hopf bundle then $p^\perp$ is homeomorphic to the complex projective plane and thus $p^\perp \setminus (\{p\} \cup q^\perp)$ is homeomorphic to $\mathbb{C}^2 \setminus \{0\}$ which is simply connected. Therefore $m(\mathfrak{G}\mathfrak{Q}) = 1$. If the space $p^\perp \setminus \{p\}$ is homeomorphic to the tangent bundle of $\mathbb{S}_2$ we obtain $m(\mathfrak{G}\mathfrak{Q}) = 2$ by [47]: 27.2 and 26.2. Thus we have $m(\mathfrak{G}\mathfrak{Q}) \in \{1, 2\}$ for a compact quadrangle with parameter 2. Moreover, a compact quadrangle with parameter 2 and its dual have different twisting numbers.

The only other possibility for a compact quadrangle with parameter s is $s = 4$ by [24]. However, up to now no example of a compact quadrangle with parameter 4 is known.

We summarise these results in the following lemma.

(2.3) Lemma: *Let p denote a point of a compact quadrangle with parameter s where $s \in \{1, 2\}$. Then $p^\perp \setminus \{p\}$ is either homeomorphic to the Möbius strip or to the cylinder or to the complex Hopf bundle or to the tangent bundle of $\mathbb{S}_2$. For the different cases the following properties hold:*

- *If $p^\perp \setminus \{p\}$ is homeomorphic to the Möbius strip, then $s = 1$ and $m(\mathfrak{G}\mathfrak{Q}) = 1$. Moreover, $p^\perp$ is homeomorphic to the real projective plane.*
- *If $p^\perp \setminus \{p\}$ is homeomorphic to the cylinder, then $s = 1$ and $m(\mathfrak{G}\mathfrak{Q}) = 0$.*
- *If $p^\perp \setminus \{p\}$ is homeomorphic to the complex Hopf bundle, then $s = 2$ and for the twisting number we have $m(\mathfrak{G}\mathfrak{Q}) = 1$. Moreover, $p^\perp$ is homeomorphic to the complex projective plane.*

– *If* $p^\perp \setminus \{p\}$ *is homeomorphic to the tangent bundle of* $\mathbb{S}_2$*, then* $s = 2$ *and for the twisting number we have* $m(\mathfrak{GQ}) = 2$.

Finally, if $\overline{\mathfrak{GQ}}$ *denotes the dual of* $\mathfrak{GQ}$*, then* $m(\overline{\mathfrak{GQ}}) \neq m(\mathfrak{GQ})$. ♋

The surprising fact is that for compact quadrangles with parameter 1 as well as for compact quadrangles with parameter 2 there is a nice interplay between topological and geometrical properties. We will see that a compact quadrangle with parameters (s,t) is antiregular if and only if either $s = t = 1$ and $m(\mathfrak{GQ}) = 0$ or $s = t = 2$ and $m(\mathfrak{GQ}) = 2$.

Remarks: The above construction works essentially also for compact polygons with parameter s. It is still an open question whether or not the twisting number together with the parameters determines the point space and line space of a compact quadrangle topologically. There is however some hope that with a tool similar to square projections this problem can be solved.

2.4 Antiregular quadrangles

We now begin with a closer examination of antiregular quadrangles. In this section we mainly recall some already known results on antiregular quadrangles. The following proposition summarises results on the topology of compact antiregular quadrangles. It also reveals the origin of the term square projection.

(2.4) Proposition: *Suppose* $\mathfrak{GQ}$ *is a compact antiregular quadrangle with parameters* (s,t)*. Then* $s = t$ *and* $s \in \{1,2\}$*. Furthermore, the map that assigns to every triad its centre is continuous.*

Moreover, if $s = 1$*, then* $m(\mathfrak{GQ}) = 0$ *and square projections are topologically equivalent to the map* $sq : \hat{\mathbb{R}} \to \hat{\mathbb{R}} : x \mapsto x^2$.

If $s = 2$*, then* $m(\mathfrak{GQ}) = 2$ *and square projections are topologically equivalent to the map* $sq : \hat{\mathbb{C}} \to \hat{\mathbb{C}} : x \mapsto x^2$.

Here $\hat{X}$ *denotes the one point compactification of* X *and the point compactifying* X *is a fixed point of the relevant map.*

Proof: By [43]: 2.3 the map that sends every triad to its centre is continuous. By [43]: 2.7 we have $s = t \in \{1,2\}$. For every point p of $\mathfrak{GQ}$ the space $p^\perp \setminus \{p\}$ is homeomorphic to the point space of a Laguerre plane of dimension $2s$ ([43]: 2.3), hence to $\mathbb{R} \times \mathbb{S}_1$ if $s = 1$ ([13]) and to the tangent bundle of $\mathbb{S}_2$ if $s = 2$ ([48]). By Lemma 2.3 this implies that $m(\mathfrak{GQ}) = 0$ if $s = 1$ and that $m(\mathfrak{GQ}) = 2$ if $s = 2$.

Let $\mu := \mu_{x,y,g}$ be a square projection. Then for every $s \in \mu(x^\perp \cap y^\perp) \subset g$ we obtain $\mu^{-1}(s) = x^\perp \cap y^\perp \cap s^\perp$, which depends continuously on s. This implies that μ is a twofold covering branched at $\pi(x, g)$ and $\pi(y, g)$. However, if $s = t \in \{1, 2\}$ then any such twofold branched covering is topologically equivalent to the respective map sq. ⚘

The above proposition implies that antiregular compact quadrangles with parameter 1 are not centric, in contrast to antiregular compact quadrangles with parameter 2 (Corollary 2.10).

(2.5) Corollary: *Suppose $\mathfrak{GQ}$ is an antiregular compact quadrangle with parameter 1, then there are three points s, t and u such that $s^\perp \cap t^\perp \cap u^\perp = \emptyset$.*

Proof: Let s and t denote two non-collinear points and let g denote a line not incident with any point of $s^\perp \cap t^\perp$. By Proposition 2.4 the square projection $\mu_{s,t,g}$ is topologically equivalent to the map $sq : \hat{\mathbb{R}} \to \hat{\mathbb{R}} : x \mapsto x^2$. In particular, the square projection is not surjective. For any point u on g not in the image of the square projection the centre $s^\perp \cap t^\perp \cap u^\perp$ is empty. ⚘

Our next aim is to prove the converse of Proposition 2.4. That is, we want to prove the following (Theorem 2.15): If $\mathfrak{GQ}$ is a compact quadrangle with parameter 2, then either $m(\mathfrak{GQ}) = 2$ and $\mathfrak{GQ}$ is centric antiregular or $m(\mathfrak{GQ}) = 1$ in which case $\mathfrak{GQ}$ is not antiregular. If $\mathfrak{GQ}$ is a compact quadrangle with parameter 1, then either $m(\mathfrak{GQ}) = 0$ and $\mathfrak{GQ}$ is antiregular or $m(\mathfrak{GQ}) = 1$ in which case $\mathfrak{GQ}$ is not antiregular. The result for compact quadrangles with parameter 1 this is already known ([37]), so we will restrict our attention to quadrangles with parameter 2.

First we recall a simple result on generalised quadrangles that we will need later.

(2.6) Lemma: *Suppose p, q and r are three points of a generalised quadrangle such that q is collinear with r but neither q nor r is collinear with p. Then $p^\perp \cap q^\perp \cap r^\perp$ contains precisely one point, namely $\pi(p, q \vee r)$.* ⚘

In general there might be different points q and r not collinear with p such that the traces $p^\perp \cap r^\perp$ and $p^\perp \cap q^\perp$ coincide. The next lemma names a condition which ascertains that two traces $p^\perp \cap q^\perp$ and $p^\perp \cap r^\perp$ coincide if and only if $q = r$.

(2.7) Lemma: *Suppose p is a point of a thick generalised quadrangle such that for any two different points $q, r \in \mathcal{P} \setminus p^{\perp}$ the set $p^{\perp} \cap q^{\perp} \cap r^{\perp}$ contains precisely one element if and only if q and r are collinear. Then for any two points $v, w \in \mathcal{P} \setminus p^{\perp}$ the sets $p^{\perp} \cap v^{\perp}$ and $p^{\perp} \cap w^{\perp}$ are equal if and only if $v = w$.*

Proof: One implication is trivial, so we only have to show that if for $v, w \in \mathcal{P} \setminus p^{\perp}$ the sets $p^{\perp} \cap v^{\perp}$ and $p^{\perp} \cap w^{\perp}$ coincide, then $v = w$. Let $z \in v^{\perp} \setminus p^{\perp}$. Then $p^{\perp} \cap v^{\perp} \cap z^{\perp} = \{\pi(p, z \vee v)\}$. Thus also $p^{\perp} \cap w^{\perp} \cap z^{\perp} = \{\pi(p, z \vee v)\}$ which, by assumption, implies $w \sim z$. Also, trivially, $\pi(p, z \vee v) \sim w$. Thus $w \in z \vee \pi(p, z \vee v) = z \vee v$. Since $\left|v^{\perp} \cap w^{\perp} \cap p^{\perp}\right| = \left|v^{\perp} \cap p^{\perp}\right| = |\mathcal{L}_p| \geq 3 \neq 1$ we conclude $v = w$. ♋

2.5 Characterisation of antiregular quadrangles

In this section we characterise all compact antiregular quadrangles. The main tool are square projections. The first step is the investigation of the connection between the twisting number of a compact quadrangle and the mapping degree of square projections. The result states that the twisting number of a compact quadrangle with parameter s equals the degree of any square projection in this quadrangle. Exploiting this result we show that, up to duality, every compact quadrangle with parameter 1 or 2 is antiregular.

(2.8) Lemma: *Suppose $\mathfrak{GQ}$ is compact quadrangle with parameter s. Then for any two points p and z and any line h_1 not incident with any point in $p^{\perp} \cap z^{\perp}$ the square projection*

$$\mu := \mu_{p,z,h_1} : p^{\perp} \cap z^{\perp} \to h_1 : x \mapsto \pi(x, h_1)$$

has mapping degree $m(\mathfrak{GQ})$.

Proof: Let $h_2 := \lambda(z, h_1)$ and let $q := \pi(z, h_1) = h_1 \cap h_2$. For $i \in \{1, 2\}$ let $x_i = \pi(p, h_i)$ and $g_i = \lambda(p, h_i)$. We now have the same situation as encountered in the preparation of the definition of the twisting number.

By assumption no point of $p^{\perp} \cap z^{\perp}$ is incident with h_1 and hence $x_1 \notin p^{\perp} \cap z^{\perp}$. Since $p^{\perp} \cap z^{\perp}$ is closed in $p^{\perp} \setminus \{p\}$ there is a neighbourhood of x_1 in $p^{\perp} \setminus \{p\}$ disjoint from $p^{\perp} \cap z^{\perp}$. By the local product structure of $p^{\perp} \setminus \{p\}$ we may choose this neighbourhood of the form $\pi(V \times G_1)$, where V is a neighbourhood of q in h_2 homeomorphic to the disc $\mathbb{D}_s$ and where G_1 is a neighbourhood of g_1 in $\mathcal{L}_p$ and a hemisphere of $\mathcal{L}_p$. Thus $G_2 := \overline{\mathcal{L}_p \setminus G_1}$, the hemisphere opposite G_1, is also homeomorphic to $\mathbb{D}_s$. Let $K := \partial G_1 = G_1 \cap G_2$, let $X_i := \{x \in p^\perp \setminus \{p\} \mid x \vee p \in G_i\}$, where $i \in \{1,2\}$, and let $Y := \{x \in p^\perp \setminus \{p\} \mid x \vee p \in K\} = X_1 \cap X_2$.

The space $p^\perp \setminus \{p\}$ is obtained by gluing together X_1 and X_2 along Y. We have seen in Section 2.3 that this gluing can be described via a homeomorphism φ on Y.

First we assume $s > 1$. Let $\kappa_i : \mathbb{R}^s \times \mathbb{D}_s \to X_i$ denote the homeomorphism induced by the coordinatisation of X_i. Let $Y_i' := \kappa_i(\mathbb{R}^s \setminus \{0\} \times \mathbb{S}_{s-1})$. Then the gluing map φ can be viewed as a map from Y_1' to Y_2'. On the level of the homotopy groups of degree $s-1$ we obtain $\varphi_* : \pi_{s-1}(Y_1') \to \pi_{s-1}(Y_2') : (a,b) \mapsto (a+m(\mathfrak{G}\mathfrak{Q})b, b)$.

Now let $Z := z^\perp \cap Y = \pi(\{z\} \times K) \cong \mathbb{S}_{s-1}$. The set Z bounds the disc $z^\perp \cap X_1^\perp = \pi(\{z\} \times G_1) \subset \pi(h_2 \setminus \{x_2, q\} \times G_1) = X_1 \setminus q^\perp$, hence Z is contractable in $X_1 \setminus q^\perp$. Thus we may assume that Z represents the element $(0,1)$ in $\pi_{s-1}(Y_1')$. But then Z represents the element $(m(\mathfrak{G}\mathfrak{Q}), 1) = \varphi_*(0,1)$ in $\pi_{s-1}(Y_2')$. The space $X_2 \setminus q^\perp$ is coordinatised by $h_1 \setminus \{x_1, q\} \times G_1$ and the restriction of μ to $Z \subset Y_2'$ is just the projection onto the first factor of this product. Thus, on the level of homotopy groups, $\mu_* : \pi_{s-1}(Y_2') \to \pi_{s-1}(h_1 \setminus \{x_1, q\}) : (a,b) \mapsto a$. Hence $\mu(Z)$ represents the element $m(\mathfrak{G}\mathfrak{Q})$ in $\pi_{s-1}(h_1 \setminus \{x_1, q\}) = \mathbb{Z}$.

By construction the preimage $\mu^{-1}(q)$ consists of only one point, namely the point x_2. Moreover, the set Z generates the homotopy group $\pi_{s-1}(z^\perp \cap X_2 \setminus \{x_2\})$ while $\mu(Z)$ represents the element $m(\mathfrak{G}\mathfrak{Q})$ in $\pi_{s-1}(h_1 \setminus \{x_1, q\})$. This implies that the map μ has mapping degree $m(\mathfrak{G}\mathfrak{Q})$ (Lemma 2.2.c).

If $s = 1$, then Z consists of two points. By definition of $m(\mathfrak{G}\mathfrak{Q})$ we have that $\mu(Z)$ is contained in one connected component of $h_1 \setminus \{x_1, q\}$ if $m(\mathfrak{G}\mathfrak{Q}) = 0$ and that $\mu(Z)$ consists of two points contained in different connected components of $h_1 \setminus \{x_1, q\}$ if $m(\mathfrak{G}\mathfrak{Q}) = 1$. This implies that the mapping degree is 0 in the first case and 1 in the second case. Thus the mapping degree equals $m(\mathfrak{G}\mathfrak{Q})$. ∎

The above lemma is the decisive step towards our goal. However, before moving the other foot, we will first note some easy but helpful consequences of the above lemma.

Since any non-surjective map on $\mathbb{S}_n$ has mapping degree 0 (Lemma 2.2.a) we immediately obtain the following result.

(2.9) Corollary: *In every compact quadrangle with parameter s and twisting number $m(\mathfrak{G}\mathfrak{Q}) > 0$ square projections are surjective.* ∎

This in turn implies:

(2.10) Corollary: *Every compact quadrangle with parameter s and twisting number $m(\mathfrak{G}\mathfrak{Q}) > 0$ is centric.*

Proof: Suppose p, q and r are three points of the quadrangle. If at least two of these points are collinear then $p^\perp \cap q^\perp \cap r^\perp$ is not empty in any quadrangle. If these points are pairwise non-collinear, then either $p^\perp \cap q^\perp = p^\perp \cap r^\perp$ in which case $p^\perp \cap q^\perp \cap r^\perp = p^\perp \cap q^\perp \neq \emptyset$ or there is a point $x \in p^\perp \cap q^\perp \setminus r^\perp$. In the latter case let $h := q \vee x$. By construction h has no point in common with $p^\perp \cap r^\perp$. By Corollary 2.9 the square projection $\mu := \mu_{p,r,h} : p^\perp \cap r^\perp \to h$ is surjective, in particular $q \in \mu(p^\perp \cap r^\perp)$. Thus $p^\perp \cap q^\perp \cap r^\perp = \mu^{-1}(q) \neq \emptyset$. ⚙

Whereas so far our considerations were rather general, we now restrict ourselves to compact quadrangles with parameter 2. Most of the following results become meaningless for compact quadrangles with parameter 1 and probably false for compact quadrangles with parameter s where $s > 2$.

(2.11) Corollary: *Suppose p, q, r are three different points of a compact quadrangle with parameter 2 where $m(\mathfrak{G}\mathfrak{Q}) \geq 2$ such that $q, r \notin p^\perp$. Then the centre $p^\perp \cap q^\perp \cap r^\perp$ consists of precisely one point if and only if q is collinear with r.*

Proof: By Lemma 2.6 we have only to prove one implication. So suppose the centre $p^\perp \cap q^\perp \cap r^\perp$ consists precisely of the point u. Let x denote a point on $p^\perp \cap q^\perp \setminus r^\perp$ and let $h := x \vee q$. Then h contains no point of $p^\perp \cap r^\perp$ and the square projection $\mu := \mu_{p,r,h} : p^\perp \cap r^\perp \to h$ has mapping degree $m(\mathfrak{G}\mathfrak{Q}) \geq 2$ by Lemma 2.8. By construction $\mu^{-1}(x) = \{\pi(r, x \vee p)\}$ and $\mu^{-1}(q) = \{u\}$. Moreover, $\mu^{-1}(\pi(r,h)) = \{\pi(p, \lambda(r,h))\}$. However, every map on $\mathbb{S}_2$ whose degree is at least 2 allows at most two points whose preimage consists of a single point (Lemma 2.2.d). Therefore, two of the points x, q and $\pi(r,h)$ are equal. Trivially, $x \neq q$ and, since $x \neq p^\perp \cap r^\perp$ by construction, also $x \neq \pi(r,h)$. Thus $q = \pi(r,h)$ and therefore $q \sim r$. ⚙

Together with Lemma 2.7 this implies the next result.

(2.12) Corollary: *Suppose p is a point of a compact quadrangle $\mathfrak{G}\mathfrak{Q}$ with parameter 2 where $m(\mathfrak{G}\mathfrak{Q}) \geq 2$. Then for any points $q, r \notin p^\perp$ we have $p^\perp \cap q^\perp = p^\perp \cap r^\perp$ if and only if $q = r$.* ⚙

For compact quadrangles with parameter 2 we can even strengthen the result of Corollary 2.9. That is, we can give a lower bound of the number of elements in the centre of a triad.

(2.13) Corollary: *Suppose $\mathfrak{G}\mathfrak{Q}$ is a compact quadrangle with parameter 2. Then the centre $p^\perp \cap q^\perp \cap r^\perp$ of any triad $\{p,q,r\}$ contains at least $m(\mathfrak{G}\mathfrak{Q})$ points.*

Proof: If $p^\perp \cap q^\perp = p^\perp \cap r^\perp$, then $p^\perp \cap q^\perp \cap r^\perp = p^\perp \cap q^\perp \cong \mathbb{S}_2$ contains infinitely many points. If $p^\perp \cap q^\perp \neq p^\perp \cap r^\perp$ choose $x \in p^\perp \cap q^\perp \setminus r^\perp$ and let $h := q \vee x$. Then no point of $p^\perp \cap r^\perp$ is incident with h and the square projection $\mu_{p,r,h} : p^\perp \cap r^\perp \to h$ has mapping degree $m(\mathfrak{G}\mathfrak{Q})$. If $m(\mathfrak{G}\mathfrak{Q}) = 0$ there is nothing to show and if $m(\mathfrak{G}\mathfrak{Q}) = 1$ the centre $p^\perp \cap q^\perp \cap r^\perp = \mu_{p,r,h}^{-1}(q)$ contains at least one point by Lemma 2.2.a. Therefore we may assume $m(\mathfrak{G}\mathfrak{Q}) \geq 2$. The two points $\pi(r,h)$ and $\pi(p,h) = x$ are two points whose primage contains one point only. By construction $r \notin \{\pi(r,h), x\}$, thus by Lemma 2.2.d we obtain $|p^\perp \cap q^\perp \cap r^\perp| = |\mu_{p,r,h}^{-1}(q)| \geq \deg(\mu_{p,r,h}) = m(\mathfrak{G}\mathfrak{Q})$. ✠

The next step is to give an upper bound of the number of points in the centre of a triad.

(2.14) Proposition: *In a compact antiregular quadrangle $\mathfrak{G}\mathfrak{Q}$ with parameter 2 whose twisting number $m(\mathfrak{G}\mathfrak{Q})$ is at least two, the centre $p^\perp \cap q^\perp \cap r^\perp$ of every triad $\{p,q,r\}$ contains at most two points.*

Proof: Suppose the statement is false. Then there exists a triad $\{p,q,r\}$ of a compact antiregular quadrangle $\mathfrak{G}\mathfrak{Q}$ with parameter 2 and twisting number $m(\mathfrak{G}\mathfrak{Q}) > 1$, such that the centre $p^\perp \cap q^\perp \cap r^\perp$ contains more than two points.

As before, our main tool are square projections from the trace $p^\perp \cap q^\perp$ to lines through r. However, unlike before, we have to consider more than one square projection. The reason is that there are maps on $\mathbb{S}_2$ of degree $m > 1$ such that two points have preimages consisting of one point and such that for at least one point the preimage consists of more than m points. Hence the investigation of just one square projection will not produce a contradiction. Our strategy is to consider a set $\{h_t \in \mathcal{L}_r \mid t \in [0,1]\}$ of lines through r such that only the line h_1 intersects $p^\perp \cap q^\perp$ and such that the set is homeomorphic to the unit interval. We then try to obtain the restriction $\mu_1 : p^\perp \cap q^\perp \setminus h_1 \to h_1 : v \mapsto \pi(v, h_1)$ as the limit of square projections from $p^\perp \cap q^\perp$ to h_t. This will lead to a contradiction near $h_1 \cap p^\perp \cap q^\perp$.

Before we investigate the square projections we first introduce some useful maps between the traces $p^\perp \cap q^\perp$ and $p^\perp \cap r^\perp$. The map

$$\zeta : p^\perp \cap q^\perp \to p^\perp \cap r^\perp : v \mapsto \pi(r, v \vee p)$$

maps the trace $p^\perp \cap q^\perp$ homeomorphically onto the trace $p^\perp \cap r^\perp$. The inverse map has the form

$$\zeta^{-1} : p^\perp \cap r^\perp \to p^\perp \cap q^\perp : w \mapsto \pi(q, w \vee p).$$

Also the map

$$\eta : p^\perp \cap q^\perp \to p^\perp \cap r^\perp : v \mapsto \pi(p, \lambda(r, v \vee q))$$

is a homeomorphism. The inverse map has the form

$$\eta^{-1} : p^\perp \cap r^\perp \to p^\perp \cap q^\perp : w \mapsto \pi(p, \lambda(q, w \vee r)).$$

Finally, let

$$\kappa := \eta^{-1} \circ \zeta : p^\perp \cap q^\perp \to p^\perp \cap q^\perp : v \mapsto \pi(p, \lambda(q, (\pi(r, v \vee p)) \vee r)).$$

Then κ is a self homeomorphism on $p^\perp \cap q^\perp$. The fixed point set of κ consists precisely of the points in $p^\perp \cap q^\perp \cap r^\perp$.

The following considerations reveal the crucial property of the map κ. Let x denote any point in $p^\perp \cap q^\perp \setminus r^\perp$ and let $h_x := \lambda(r, p \vee x) = \zeta(x) \vee r$. Consider the square projection $\mu_x := \mu_{p,q,h_x} : p^\perp \cap q^\perp \to h_x$. By Lemma 2.8 we know that the mapping degree of μ_x is $m(\mathfrak{G}\mathfrak{Q})$. By construction, $\zeta(x)$ is collinear with p, thus $\zeta(x) = \pi(p, h_x)$. Since the preimage of $\pi(p, h_x)$ under the square projection μ_x contains only one element we obtain $\mu_x^{-1}(\mu_x(x)) = \mu_x^{-1}(\zeta(x)) = \{x\}$. Our next task is to find the other point y in $p^\perp \cap q^\perp$ with $\mu_x^{-1}(\mu_x(y)) = \{y\}$. We claim that $\kappa(x)$ is this point. By definition

$$\begin{aligned}
\kappa(x) &= \eta^{-1}(\zeta(x)) \\
&= \pi(p, \lambda(q, \zeta(x) \vee r)) \\
&= \pi(p, \lambda(q, h_x)) \\
&= \pi(p, \pi(q, h_x) \vee q)
\end{aligned}$$

which implies

$$\mu_x(\kappa(x)) = \pi(\pi(p, \pi(q, h_x) \vee q), h_x) = \pi(q, h_x).$$

Since the preimage of $\pi(q, h_x)$ under the square projection μ_x consists of only one element we obtain $\mu_x^{-1}(\mu_x(\kappa(x))) = \{\kappa(x)\}$ as claimed. Together with the above this shows that x and $\kappa(x)$ are two points in $p^\perp \cap q^\perp$ such that $\mu_x^{-1}(\mu_x(x)) = \{x\}$ and $\mu_x^{-1}(\mu_x(\kappa(x))) = \{\kappa(x)\}$. By Lemma 2.2.d these are the only two points y in the trace $p^\perp \cap q^\perp$ such that $\mu_x^{-1}(\mu_x(y))) = \{y\}$.

Now, with notation as above, the square projection $\mu_x : p^\perp \cap q^\perp \to h_x$ has mapping degree $m = m(\mathfrak{G}\mathfrak{Q}) \geq 2$ (Lemma 2.8). By assumption $\mu_x^{-1}(r) = p^\perp \cap q^\perp \cap r^\perp$ contains at least three points. By the above $\mu_x(x)$ and $\mu_x(\kappa(x))$ are the only two points in h_x whose preimage consists of only one point. Thus there exists a map

$$\varphi : [0, 2] \to p^\perp \cap q^\perp \setminus \{x, \kappa(x)\}$$

which is injective on $[0, 2[$ such that $\varphi(0) = \varphi(2) \in \mu_x^{-1}(r)$ and also $\varphi(1) \in \mu_x^{-1}(r)$, such that φ represents a generating element of the fundamental group of the space $p^\perp \cap q^\perp \setminus \{x, \kappa(x)\}$ and such that the restriction of $\mu_x \circ \varphi$ to $[0, 1]$ as well as the restriction of $\mu_x \circ \varphi$ to $[1, 2]$ represents a nontrivial element in the fundamental group $\pi_1(h_x \setminus \{\mu_x(x), \mu_x(\kappa(x))\})$. This implies that $\mu_x(x)$ and $\mu_x(\kappa(x))$ lie in different connected components of $h_x \setminus \mu_x(\varphi([0, 1]))$ as well as in different connected components of $h_x \setminus \mu_x(\varphi([1, 2]))$. It also implies that x and $\kappa(x)$ are contained in different connected components of $p^\perp \cap q^\perp \setminus \varphi([0, 2])$. Since $p^\perp \cap q^\perp \cap r^\perp$ contains at least three points, there is a point $z \in p^\perp \cap q^\perp \cap r^\perp$ different from $\varphi(0)$ and different from $\varphi(1)$. We may choose φ in such a way that z is contained in the same connected component of $p^\perp \cap q^\perp \setminus \varphi([0, 2])$ as x. In particular, $z \notin \varphi[0, 2]$. For the construction of the map φ the assumption that the twisting number of the quadrangle and hence the mapping degree of square projections is at least two as well as the assumption that the centre $p^\perp \cap q^\perp \cap r^\perp$ consists of more than two points is essential. Let $\Sigma := \varphi[0, 2]$.

By construction x and z are contained in a common connected component of the space $p^\perp \cap q^\perp \setminus \Sigma$, thus there exists a path $\rho : [0, 1] \to p^\perp \cap q^\perp$ from x to z missing Σ. Replacing z by $\rho(\tau)$ where $\tau := \min\{t \in [0, 1] \mid \rho(t) \in r^\perp\}$ if necessary, we may assume $\rho([0, 1[) \cap r^\perp = \emptyset$. Let $\psi := \kappa \circ \rho$. Then ψ is a path from $\kappa(x)$ to $z = \kappa(z)$. Since $\rho([0, 1[) \cap r^\perp = \emptyset$ we have $\rho(t) \neq \psi(t)$ for all $t \in [0, 1[$ as well as $\psi([0, 1[) \cap r^\perp = \emptyset$. Since z and $\kappa(x)$ lie in different connected components of $p^\perp \cap q^\perp \setminus \Sigma$ there is some $\tau \in [0, 1]$ such that $\psi(\tau) \in \Sigma$ but $\psi(t) \notin \Sigma$ for all $t \in]\tau, 1]$. Replacing the map φ by the map $t \mapsto \varphi(2 - t)$ if necessary we may assume $\psi(\tau) \in \varphi([1, 2])$. Moreover, since $\{\varphi(1), \varphi(2)\} \in p^\perp \cap q^\perp \cap r^\perp \setminus \{z\}$ and $\psi([0, 1]) \cap r^\perp = \{z\}$ we obtain $\psi(\tau) \in \varphi(]1, 2[)$, i.e., $\psi(\tau) \notin \varphi([0, 1])$. There might however exist some $t \in]0, \tau[$ such that $\psi(t) \in \varphi([0, 1])$. Fortunately, this is of no great harm. Consider the map

$$\Phi : [0, 2] \times (\{0\} \cup [\tau, 1]) \cup \{0, 1, 2\} \times [0, \tau] \to p^\perp \cap q^\perp : (t, s) \mapsto \varphi(t).$$

This map can be extended to a map

$$\Phi : [0,2] \times [0,1] \to p^{\perp} \cap q^{\perp}$$

such that for every $s \in [0, \tau[$ neither $\rho(s)$ nor $\psi(s)$ is contained in $\Phi([0,2] \times \{s\})$. In particular this implies that for every $s \in [0,1]$ neither $\rho(s)$ nor $\psi(s)$ is contained in $\Phi([0,1] \times \{s\})$.

For $t \in [0,1]$ let $h_t := \lambda(r, p \vee \rho(t)) = \zeta(\rho(t)) \vee r$. For $t \in [0,1[$ let μ_t denote the square projection $\mu_{p,q,h_t} : p^{\perp} \cap q^{\perp} \to h_t$. Since $\rho(0) = x$ we have $h_0 = h_x$ and $\mu_0 = \mu_x$. Moreover, by considerations made above we know that $\rho(t)$ and $\psi(t) = \kappa(\rho(t))$ are the two points in $p^{\perp} \cap q^{\perp}$ whose images under μ_t have only one preimage under μ_t. That is, we have $\mu_t^{-1}(\mu_t(\rho(t))) = \{\rho(t)\}$ and $\mu_t^{-1}(\mu_t(\psi(t))) = \{\psi(t)\}$, hence also $\mu_t(\rho(t)) = \pi(p, h_t) = \pi(p, \zeta(\rho(t)) \vee r)$ and $\mu_t(\psi(t)) = \pi(q, h_t) = \pi(q, \zeta(\rho(t)) \vee r)$.

Since $\rho(t) \notin r^{\perp}$ for $t \in [0,1[$, the lines h_t do not intersect the trace $p^{\perp} \cap q^{\perp}$ and the square projections μ_t do really exist. For $t = 1$ the situation is different. We have $\rho(1) = z \in r^{\perp}$ and thus $h_1 = z \vee r$, hence $h_1 \cap p^{\perp} \cap q^{\perp} = \{z\}$. Let $\mu_1 : p^{\perp} \cap q^{\perp} \setminus \{z\} \to h_1 : v \mapsto \pi(v, h_1)$. This map is not a square projection. We also do not know whether or not it can be extended to a continuous map on the whole of $p^{\perp} \cap q^{\perp}$.

Next consider the space

$$X := \{(v,t) \in r^{\perp} \times [0,1] \mid t \in [0,1] \text{ and } v \in \mathcal{P}_{h_t} = \mu_t(p^{\perp} \cap q^{\perp})\}.$$

Since h_t depends continuously on t we conclude that the space X is homeomorphic to $\mathbb{S}_2 \times [0,1]$. Let

$$Y := \{(v,t) \in r^{\perp} \times [0,1] \mid t \in [0,1] \text{ and } v \in \mu_t(\Phi([0,1] \times \{t\}))\}.$$

By construction $\mu_t(\Phi(0,t)) = \mu_t(\Phi(1,t)) = \mu_t(\varphi(0)) = r$ for $t \in [0,1]$, thus we can view $Y \subset X$ as a continuous image of $\mathbb{S}_1 \times [0,1]$. Since $\Phi([0,1] \times \{0\}) = \varphi([0,1])$ separates h_0 into at least two connected components, one of which contains $\mu_0(x)$ and $\mu_0(z) = r$ but not $\mu_0(\kappa(x))$ we obtain that $X \setminus Y$ consists of at least two connected components, one of which contains $\mu_0(x)$ and $\mu_0(z) = r$ but not $\mu_0(\kappa(x))$. We have seen above that $\mu_t^{-1}(\mu_t(\rho(t))) = \{\rho(t)\}$ and $\mu_t^{-1}(\mu_t(\psi(t))) = \{\psi(t)\}$ and that neither $\rho(t)$ nor $\psi(t)$ are contained in $\Phi([0,1] \times \{t\})$. Thus the sets $A := \{(\mu_t(\rho(t)), t) \mid t \in [0,1[\}$ and $B := \{(\mu_t(\psi(t)), t) \mid t \in [0,1[\}$ are contained in different connected components of $X \setminus Y$. Let $t_n \in [0,1[$ denote a sequence converging to 1. Let $w_n := \mu_{t_n}(\psi(t_n)) \in h_{t_n}$. By compactness of X we may assume that there is some $w \in h_1$ such that $\lim_{n\to\infty} w_n = w$. Since $(w_n, t_n) \in B$ we know that $(w, 1)$ is contained in the closure of the connected component of $X \setminus Y$ containing B.

On the other hand we have

$$w_n = \mu_{t_n}(\psi(t_n)) = \pi(q, h_{t_n})$$

and thus, by continuity of π we obtain

$$w = \lim_{n\to\infty} w_n = \lim_{n\to\infty} \pi(q, h_{t_n}) = \pi(q, h_1) = \pi(q, z \vee r) = z.$$

An analogous argument shows that $z = \pi(p, h_1) = \lim_{n\to\infty} \mu_{t_n}(\rho(t_n))$. Thus $(w, 1) = (z, 1)$ is also in the closure of the component of $X \setminus Y$ containing A. Together with the above this implies that $(w, 1)$ is in the closure of two different connected components of $X \setminus Y$. This implies $(w, 1) = (z, 1) \in Y$, hence $w = z \in \mu_1(\varphi([0, 1]))$. However, no element of $\varphi([0, 1]) \subset p^\perp \cap q^\perp \setminus \{z\}$ is collinear with z. Thus $w = z \in \mu_1(\varphi([0, 1])) = \pi(\varphi([0, 1]) \times \{h_1\})$ cannot hold. This contradiction completes the proof. ⚘

Remark: The above lemma also shows again that the twisting number of a compact quadrangle with parameter 2 is at most two. For compact quadrangles with parameter s whose twisting number equals 1 there exists no upper bound for the number of elements in the centre of a triad. In regular quadrangles for example there are two types of centres. Either a centre consists of precisely one point or a centre is isomorphic to a line pencil ([45]).

We are now in a position to state the main result of this section, a characterisation of compact antiregular quadrangles.

(2.15) Theorem: *Suppose $\mathfrak{G}\mathfrak{Q}$ is a compact quadrangle with topological parameter 2. Then either $m(\mathfrak{G}\mathfrak{Q}) = 2$ and $\mathfrak{G}\mathfrak{Q}$ is centric antiregular or $m(\mathfrak{G}\mathfrak{Q}) = 1$ in which case $\mathfrak{G}\mathfrak{Q}$ is not antiregular.*

Suppose $\mathfrak{G}\mathfrak{Q}$ is a compact quadrangle with topological parameter 1. Then either $m(\mathfrak{G}\mathfrak{Q}) = 0$ and $\mathfrak{G}\mathfrak{Q}$ is antiregular or $m(\mathfrak{G}\mathfrak{Q}) = 1$ in which case $\mathfrak{G}\mathfrak{Q}$ is not antiregular.

A compact quadrangle with parameters $(s, t) \notin \{(1, 1), (2, 2)\}$ is not antiregular.

Proof: By Lemma 2.3, we know $m(\mathfrak{G}\mathfrak{Q})) \in \{1, 2\}$. If $m(\mathfrak{G}\mathfrak{Q}) = 1$ then $\mathfrak{G}\mathfrak{Q}$ is not antiregular by Proposition 2.4. If $m(\mathfrak{G}\mathfrak{Q}) = 2$ then the centre $p^\perp \cap q^\perp \cap r^\perp$ of every triad $\{p, q, r\}$ contains at least two points by Corollary 2.13 and at most two points by Proposition 2.14.

For $s = 1$ the result was shown in [37] and [39]. ⚘

With Lemma 2.3 we immediately obtain the following result.

(2.16) Corollary: *Suppose $\mathfrak{GQ}$ is a compact quadrangle with parameter s where $s \in \{1, 2\}$. Then either $\mathfrak{GQ}$ or the dual quadrangle is antiregular.* ♋

For compact quadrangles with parameter 1, there is a ready way to determine antiregularity.

(2.17) Lemma: *A compact quadrangle with parameter 1 is antiregular if and only if there are three points s, t and u such that $s^\perp \cap t^\perp \cap u^\perp = \emptyset$.*

Proof: Suppose there are three necessarily pairwise non-collinear points s, t and u whose centre $s^\perp \cap t^\perp \cap u^\perp$ is empty. Let g denote a line incident with u that is not incident with any point of $s^\perp \cap t^\perp$. The square projection $\mu_{s,t,g}$ is not surjective, thus its mapping degree vanishes (Lemma 2.2.a). This implies that the twisting number of the quadrangle vanishes also. Thus, by Theorem 2.15, the quadrangle is antiregular.

The converse implication holds by Corollary 2.5. ♋

3 Laguerre planes and antiregular quadrangles

Introduction

We now come to the connection between generalised quadrangles and circle planes. This connection is strongest for Laguerre planes. In this chapter we therefore study the connection between compact antiregular quadrangles and finite-dimensional locally compact connected Laguerre planes. All of the results presented in this chapter have already appeared elsewhere. The connection between flat Laguerre planes and generalised quadrangles was investigated in my Diplomarbeit [36] and published in [37]. The relationship between locally compact Laguerre planes with parameter 2 and generalised quadrangles was first studied in my Ph.D. thesis [39] and published in [41], [42] and [43].

To be able to compare this situation with the connection of Möbius planes or Minkowski planes and generalised quadrangles, we give some indication of the proofs. For one result, namely that the quadrangle constructed from a locally compact Laguerre plane with parameter 2 is antiregular, a new proof is provided. The new proof is considerably shorter than the older one presented in [42] and takes advantage of the facts supplied in the previous chapter.

The chapter is organised as follows. In the first section we will show that a certain derived structure of an antiregular generalised quadrangle yields a Laguerre plane. In the second section the Lie geometry of a Laguerre plane is introduced. It is shown that for finite-dimensional connected Laguere planes the Lie geometry is a generalised quadrangle. In the final section we define natural topologies on the Lie geometry of a finite-dimensional connected Laguere plane and show that this yields a compact generalised quadrangle. We also see that the generalised quadrangle obtained from a Laguerre plane is antiregular and that the derivation is inverse to the construction of the Lie geometry.

3.1 Laguerre planes constructed from antiregular quadrangles

In Chapter 2 we have already exhibited the connection between a certain quadric Q in projective 4-space and the classical Laguerre plane. We calculated that for every point $p \in Q$ the incidence structure $\mathfrak{Cp}(Q,p) = (p^\perp \setminus \{p\}, \{p^\perp \cap q^\perp \mid q \in Q \setminus p^\perp\}, \in)$ is a Laguerre plane. We have also convinced ourselves that the points of this quadric and the lines of the projective space which are contained entirely in Q form a compact antiregular quadrangle with parameter 1. The same construction works also in the complex case and establishes a natural relationship between the classical locally compact Laguerre plane with parameter 2 and a certain compact antiregular quadrangle with parameter 2. So, one might conjecture that the above construction works for antiregular quadrangles in general. And indeed, given a point of a thick antiregular quadrangle, one can construct a Laguerre plane.

(3.1) Theorem: *Suppose $\mathfrak{GQ} = (\mathcal{P}, \mathcal{L})$ is a thick antiregular quadrangle. Then for every point $p \in \mathcal{P}$ the derived structure*

$$\mathfrak{L}(\mathfrak{GQ}, p) := (p^\perp \setminus \{p\}, \{p^\perp \cap q^\perp \mid q \in \mathcal{P} \setminus p^\perp\}, \{\mathcal{P}_l \setminus \{p\} \mid l \in \mathcal{L}_p\}, \in)$$

is a Laguerre plane.

Proof: Let $\mathcal{S} := p^\perp \setminus \{p\}$ and let $\mathcal{K} := \{p^\perp \cap q^\perp \mid q \in \mathcal{P} \setminus p^\perp\}$. In addition, let $\iota : \mathcal{P} \setminus \{p\} \to \mathcal{S} \cup \mathcal{K}$ denote the map that is the identity on $p^\perp \setminus \{p\}$ and maps $q \in \mathcal{P} \setminus p^\perp$ to the trace $p^\perp \cap q^\perp$.

If s, t, u are three pairwise non-parallel points of $\mathcal{S}$, then $\{\iota^{-1}(s), \iota^{-1}(t), \iota^{-1}(u)\}$ forms a triad. By construction the centre $\iota^{-1}(s)^\perp \cap \iota^{-1}(t)^\perp \cap \iota^{-1}(u)^\perp$ contains p, thus, by antiregularity, there is precisely one further point q in $\iota^{-1}(s)^\perp \cap \iota^{-1}(t)^\perp \cap \iota^{-1}(u)^\perp$. Therefore $\alpha(s,t,u) := \iota(q) = \iota((\iota^{-1}(s)^\perp \cap \iota^{-1}(t)^\perp \cap \iota^{-1}(u)^\perp) \setminus \{p\}) \in \mathcal{K}$ is the unique circle incident with s, t and u. This also implies that the restriction $\iota : \mathcal{P} \setminus p^\perp \to \mathcal{K}$ is bijective.

If $(s,k) \in \mathcal{S} \times \mathcal{K}$ is a flag and if t is a point not parallel to s, then $\beta(s,k,t) := \iota\big(\pi(\iota^{-1}(t), \iota^{-1}(s) \vee \iota^{-1}(k))\big)$ is the unique circle through t touching k in s.

If $(s,k) \in \mathcal{S} \times \mathcal{K}$ then $\iota(\pi(\iota(k), \iota^{-1}(s) \vee p))$ is the unique point on k parallel to s.

Finally, thickness of the quadrangle implies that the derived structure is not degenerate. ♋

Next suppose the quadrangle given is a topological quadrangle. By requiring the restrictions $\iota : p^\perp \setminus \{p\} \to \mathcal{S}$ and $\iota : \mathcal{P} \setminus p^\perp \to \mathcal{K}$ to be homeomorphisms, the sets $\mathcal{S}$ and $\mathcal{K}$ are equipped with topologies. With these topologies the maps β and γ become continuous since they can be expressed entirely in terms of ι, π and λ. However, the map α and intersection of circles might not be continuous. To make sure that these operations are also continuous, one has to impose further restrictions on the quadrangle. If the quadrangle $\mathfrak{GQ}$ is a compact connected antiregular quadrangle, then, by Proposition 2.4, the parameter of $\mathfrak{GQ}$ is 1 or 2. This implies that the map that sends every triad with non-empty centre to its centre is continuous ([43] or Proposition 2.4). But if the map that assigns to every triad its centre is continuous, then for every point p of the quadrangle the map α and intersection of circles are continuous in the Laguerre plane $\mathfrak{L}(\mathfrak{GQ}, p)$.

Finally, if $\mathfrak{GQ}$ is a compact quadrangle with parameter s, then for any two non-collinear points p, q of the quadrangle, the trace $p^\perp \cap q^\perp$ is homeomorphic to $\mathcal{L}_q$, hence to $\mathbb{S}_s$. Thus circles of the Laguerre plane $\mathfrak{L}(\mathfrak{GQ}, p)$ are homeomorphic to $\mathbb{S}_s$. The point space of the derived Laguerre plane is homeomorphic to $\mathcal{P} \setminus p^\perp$, hence connected and locally compact. Therefore we obtain the following result.

(3.2) Theorem: *Suppose $\mathfrak{GQ} = (\mathcal{P}, \mathcal{L})$ is a compact antiregular quadrangle with parameter s. Then $s \in \{1, 2\}$ and for every point $p \in \mathcal{P}$ the derived structure*

$$\mathfrak{L}(\mathfrak{GQ}, p) = \left(p^\perp \setminus \{p\}, \{p^\perp \cap q^\perp \mid q \in \mathcal{P} \setminus p^\perp\}, \{\mathcal{P}_l \setminus \{p\} \mid l \in \mathcal{L}_p\}, \in\right)$$

is a locally compact Laguerre with parameter s. ♋

Remark: In the finite case one obtains by the same reasoning a strikingly similar result. Suppose $\mathfrak{GQ} = (\mathcal{P}, \mathcal{L})$ is a thick finite antiregular quadrangle with parameter s, i.e., a generalised quadrangle where every point is incident with $s+1$ lines and every line is incident with $s+1$ points. Then for every point $p \in \mathcal{P}$ the derived structure $\mathfrak{L}(\mathfrak{GQ}, p)$ is a finite Laguerre with parameter s, i.e., a Laguerre plane where every circle is incident with $s+1$ points.

3.2 Antiregular quadrangles constructed from Laguerre planes

The next question is whether or not each locally compact connected Laguerre plane can be obtained as the derived structure of a compact antiregular quadrangle. Or, to put it in other words, can one construct a generalised quadrangle from any Laguerre plane?

Suppose $\mathfrak{L} = (\mathcal{S}, \mathcal{K}, \mathcal{G})$ is a Laguerre plane with point set $\mathcal{S}$, set of circles $\mathcal{K}$ and set of equivalence classes $\mathcal{G}$. Recall that $\overline{\mathcal{B}}$ denotes the set of extended tangent pencils, where an extended tangent pencil is a tangent pencil together with its support. Let $\infty_{\mathfrak{L}}$ denote an additional point. An **extended equivalence class** of the Laguerre plane is an equivalence class together with the point $\infty_{\mathfrak{L}}$. Let $\overline{\mathcal{G}}$ denote the set of extended equivalence classes. The **Lie geometry $\mathfrak{Lie}(\mathfrak{L})$ of a Laguerre plane $\mathfrak{L} = (\mathcal{S}, \mathcal{K}, \mathcal{G})$** is defined as follows:

$$\mathfrak{Lie}(\mathfrak{L}) := (\mathcal{P} := \mathcal{S} \cup \mathcal{K} \cup \{\infty_{\mathfrak{L}}\}, \mathcal{L} := \overline{\mathcal{B}} \cup \overline{\mathcal{G}}, \in) .$$

So points of the Lie geometry are points and circles of the Laguerre plane plus one extra point. Two points of the Lie geometry are collinear, if, in the Laguerre plane, they correspond to a pair of parallel points, to a pair of tangent circles or to a flag. The additional point $\infty_{\mathfrak{L}}$ is collinear with all points of the Laguerre plane but not with any circle.

We will see below hat for any non-incident point-line pair $p, g \in \mathcal{P} \times \mathcal{L}$ there exists precisely one point $\pi(p, g)$ on g collinear with p and precisely one line $\lambda(p, g)$ through p intersecting g — unless $p \in \mathcal{K}$ and $g \in \overline{\mathcal{B}}$ such that the support $\sigma(g)$ is not incident with p. Again, one has to impose additional requirements on the Laguerre plane to make sure that even in this last case there exists precisely one point $\pi(p, g)$ on g collinear with p and precisely one line $\lambda(p, g)$ through p intersecting g.

In the finite case one requires the Laguerre plane to be of odd order ([30]). In the topological case we require the Laguerre plane $\mathfrak{L}$ to be finite-dimensional and connected.

(3.3) Lemma: *Suppose p is a circle and l is a tangent pencil of a finite-dimensional connected Laguerre plane such that the support of l is not incident with p. Then precisely one circle in l touches p.*

Proof: The trick is to transport the situation to the derived projective plane $\mathfrak{P}(\mathfrak{L}, \sigma(l))$ as defined in Section 1.3. In the derived plane the circle p becomes an oval $\overline{p}$ tangent to the line at infinity l_∞. The tangent pencil l becomes a line pencil through a point u on l_∞ where $u \notin \overline{p}$. Since $\overline{p}$ is an oval, there exists precisely one other line $\overline{q}$ through u that is tangent to $\overline{p}$ (Theorem 1.1). The circle q of the Laguerre plane corresponding to $\overline{q}$ is the unique circle in the tangent pencil l touching p. ✠

The crucial point in the above proof is the fact, that the tangents of the oval $\overline{p}$ form an oval in the dual plane. We now have gathered enough information to prove the next result.

(3.4) Theorem: *The Lie geometry $\mathfrak{Lie}(\mathfrak{L})$ of a finite-dimensional connected Laguerre plane is a generalised quadrangle.*

Proof: We have to show that for every point $p \in \mathcal{P} = \mathcal{S} \cup \mathcal{K} \cup \{\infty_{\mathfrak{L}}\}$ and every line $l \in \overline{\mathcal{B}} \cup \overline{\mathcal{G}}$ not incident with p there exists precisely one line $\lambda(p,l)$ through p meeting l. The point of intersection then is $\pi(p,l)$. Thickness of the quadrangle is obvious.

We have to consider the following different possibilities for the pair $(p,l) \in \mathcal{P} \times \mathcal{L}$ with $p \notin l$:

- If $p = \infty_{\mathfrak{L}}$ and $l \in \overline{\mathcal{B}}$, then $\pi(p,l) = \sigma(l)$ and $\lambda(p,l)$ is the extended parallel class of $\sigma(l)$.
- If $p \in \mathcal{S}$ and $l \in \overline{\mathcal{G}} \setminus \Gamma(p)$, then $\pi(p,l) = \infty_{\mathfrak{L}}$ and $\lambda(p,l)$ is the extended parallel class of p.
- If $p \in \mathcal{S}$ and $l \in \overline{\mathcal{B}}$ with $\sigma(l) \neq p$, then there are two cases. If p is parallel to $\sigma(l)$, then $\pi(p,l) = \sigma(l)$ and $\lambda(p,l) = \Gamma(\sigma(l)) = \Gamma(p)$. If p is not parallel to $\sigma(l)$, then $\pi(p,l) = \beta(\sigma(l),k,p)$, where k is any circle of l and $\lambda(p,l) = \overline{B}(p,\pi(p,l))$.
- If $p \in \mathcal{K}$ and $l \in \overline{\mathcal{G}}$, then $\pi(p,l) = p \cap l$ and $\lambda(p,l) = \overline{B}(\pi(p,l),p)$.
- If $p \in \mathcal{K}$ and $l \in \overline{\mathcal{B}}$ with $p \notin l$, then there are two cases. If p is incident with $\sigma(l)$, then $\pi(p,l) = \sigma(l)$ and $\lambda(p,l) = \overline{B}(\sigma(l),p)$. If p is not incident with $\sigma(l)$, then, by the preceding lemma, there exists precisely one circle $q \in l$ touching p, hence $\pi(p,l) = q$ and $\lambda(p,l) = \overline{B}((p \cap q),p)$. ◈

Remark: Note that only in the very last case the topological requirements are needed.

Notation: Now that we know that the Lie geometry of a finite-dimensional connected Laguerre plane is a generalised quadrangle, we denote the Lie geometry of a finite-dimensional connected Laguerre plane $\mathfrak{L}$ by $\mathfrak{GQ}(\mathfrak{L})$.

3.3 Constructing topologies on the Lie geometry

In the previous section we have seen that the Lie geometry of a finite-dimensional connected Laguerre plane is a generalised quadrangle. Of course, starting with a topological Laguerre plane we want to obtain a topological quadrangle. There are already topologies on the two subsets of the point space corresponding to the point space and circle space of the Laguerre plane. There is also a natural way to define topologies on certain subsets of the line space. These topologies are such, that certain natural restrictions of the maps π and λ are continuous. The difficult part is to define a topology on the whole point space and a topology on the whole line space respecting the given topologies, such that the quadrangle becomes a topological quadrangle.

To define a topology on the whole point space requires, amongst other things, a rule to decide when a sequence of circles converges to a point. The first try might be to consider circles as closed subsets of the point space and look at the limit of the sequence of closed subsets. In the flat case this works. There the situation is as follows: A sequence of circles converges to a point s if and only if the circles, considered as subsets, basically converge to a closed interval bounded by s and $\infty_{\mathfrak{L}}$ of the extended parallel class of s. This approach is taken in [11] and [41]. Alas, for Laguerre planes with parameter 2 this approach fails. The reason is that if circles, considered a subsets, converge to a subset A of an extended parallel class, then A is already the whole parallel class and there is no easy way to single out a point in this parallel class, to which the sequence of circles might converge. However, the point can be singled out if not just the sequence of circles, but also a sequence of tangent pencils is considered.

The main result of this second approach will be cited below. Our aim in this section is to see that we meet all requirements stated in the cited theorem. Before we start with the nitty gritty we introduce the concept of the Hausdorff metric and some new, convenient notation.

A practical tool when one wants to put a topology on a set of closed subsets of a metric space is the Hausdorff metric on the space of closed subsets of a metric space. For the precise definition of the Hausdorff metric we refer to [29]: Section 0. All we use is the following characterisation of convergence: A sequence of closed non-empty sets A_i converges to a closed set $A \neq \emptyset$ with respect to the Hausdorff metric, if and only if every convergent sequence of points $a_i \in A_i$ converges to some $a \in A$ and if every point in $a \in A$ is the limit point of a sequence of points $a_i \in A_i$ ([29]: 0.6). An immediate consequence of this characterisation is that any given metric space X is naturally homeomorphic to $\{\{x\} \mid x \in X\}$ equipped with the topology induced by the Hausdorff metric.

We now often work in the generalised quadrangle as well as in the Laguerre plane. To avoid confusion, the flag space of a Laguerre plane will be denoted $\mathcal{F}_{\mathfrak{L}}$ and the flag space of a generalised quadrangle will be denoted $\mathcal{F}_{\mathfrak{GQ}}$. The combinatorial metric on the incidence graph of a generalised quadrangle $\mathfrak{GQ} = (\mathcal{P}, \mathcal{L})$ induces the following notation. For $v \in \mathcal{P} \cup \mathcal{L}$ and $i = 1, 2, 3, 4$ the set of points or lines at distance i from v is denoted $\boldsymbol{\Delta_i(v)}$. So, for a point $p \in \mathcal{P}$ we obtain:

$$\Delta_1(p) = \mathcal{L}_p, \qquad \Delta_2(p) = p^\perp \setminus \{p\}, \qquad \Delta_3(p) = \mathcal{L} \setminus \Delta_1(p), \qquad \Delta_4(p) = \mathcal{P} \setminus p^\perp.$$

For a line $l \in \mathcal{L}$ we obtain:

$$\Delta_1(l) = \mathcal{P}_l, \qquad \Delta_2(l) = l^\perp \setminus \{l\}, \qquad \Delta_3(l) = \mathcal{P} \setminus \Delta_1(l), \qquad \Delta_4(l) = \mathcal{L} \setminus l^\perp.$$

Suppose the generalised quadrangle $\mathfrak{GQ}$ is the Lie geometry of a Laguerre plane $\mathfrak{L}$, i.e.; $\mathfrak{GQ} = \mathfrak{GQ}(\mathfrak{L})$. If we take $p = \infty_{\mathfrak{L}}$ we obtain the following identifications:

$$\Delta_1(\infty_{\mathfrak{L}}) = \overline{\mathcal{G}}, \qquad \Delta_2(\infty_{\mathfrak{L}}) = \mathcal{S}, \qquad \Delta_3(\infty_{\mathfrak{L}}) = \overline{\mathcal{B}}, \qquad \Delta_4(\infty_{\mathfrak{L}}) = \mathcal{K}.$$

In the proof to Theorem 3.4 we have seen that viewed in the Laguerre plane, the maps π and λ have different geometric meanings, depending on the type of p and l. We therefore define the following restrictions of the maps π and λ:

$$\begin{aligned}
&\pi_{0,3} : \Delta_3(\infty_{\mathfrak{L}}) \to \Delta_2(\infty_{\mathfrak{L}}) : l \mapsto \pi(\infty_{\mathfrak{L}}, l)\\
&\lambda_{0,3} : \Delta_3(\infty_{\mathfrak{L}}) \to \Delta_1(\infty_{\mathfrak{L}}) : l \mapsto \lambda(\infty_{\mathfrak{L}}, l)\\
&\pi_{2,3} : \Delta_2(\infty_{\mathfrak{L}}) \times \Delta_3(\infty_{\mathfrak{L}}) \setminus \left(\mathcal{F}_{\mathfrak{GQ}} \cup \pi^{-1}(\Delta_2(\infty_{\mathfrak{L}}))\right) \to \Delta_4(\infty_{\mathfrak{L}}) : (p, l) \mapsto \pi(p, l)\\
&\lambda_{2,3} : \Delta_2(\infty_{\mathfrak{L}}) \times \Delta_3(\infty_{\mathfrak{L}}) \setminus \left(\mathcal{F}_{\mathfrak{GQ}} \cup \pi^{-1}(\Delta_2(\infty_{\mathfrak{L}}))\right) \to \Delta_3(\infty_{\mathfrak{L}}) : (p, l) \mapsto \lambda(p, l)\\
&\pi_{4,1} : \Delta_4(\infty_{\mathfrak{L}}) \times \Delta_1(\infty_{\mathfrak{L}}) \to \Delta_2(\infty_{\mathfrak{L}}) : (p, l) \mapsto \pi(p, l)\\
&\lambda_{4,1} : \Delta_4(\infty_{\mathfrak{L}}) \times \Delta_1(\infty_{\mathfrak{L}}) \to \Delta_3(\infty_{\mathfrak{L}}) : (p, l) \mapsto \lambda(p, l)\\
&\pi_{4,3} : \Delta_4(\infty_{\mathfrak{L}}) \times \Delta_3(\infty_{\mathfrak{L}}) \setminus \left(\mathcal{F}_{\mathfrak{GQ}} \cup \pi^{-1}(\Delta_2(\infty_{\mathfrak{L}}))\right) \to \Delta_4(\infty_{\mathfrak{L}}) : (p, l) \mapsto \pi(p, l)\\
&\lambda_{4,3} : \Delta_4(\infty_{\mathfrak{L}}) \times \Delta_3(\infty_{\mathfrak{L}}) \setminus \left(\mathcal{F}_{\mathfrak{GQ}} \cup \pi^{-1}(\Delta_2(\infty_{\mathfrak{L}}))\right) \to \Delta_3(\infty_{\mathfrak{L}}) : (p, l) \mapsto \lambda(p, l)
\end{aligned}$$

Finally, as a disguised restriction of the map λ we consider the map

$$\psi : \Delta_2(\infty_{\mathfrak{L}}) \to \Delta_1(\infty_{\mathfrak{L}}) : p \mapsto p \vee \infty_{\mathfrak{L}}.$$

We now have all notation to cite the main result of [41].

(3.5) Theorem: *Suppose* $\infty \in \mathcal{P}$ *is a point of a generalised quadrangle* $\mathfrak{GQ} = (\mathcal{P}, \mathcal{L}, \mathcal{F}_{\mathfrak{GQ}})$ *such that:*

- *the sets* $\Delta_i(\infty)$*, where* $i = 1, 2, 3, 4$*, are topological spaces,*
- *the space* $\Delta_1(\infty)$ *is homeomorphic to* $\mathbb{S}_s$*,*
- *the spaces* $\Delta_2(\infty)$ *and* $\Delta_4(\infty)$ *are metric spaces, such that for every line* l *in* $\Delta_1(\infty)$ *resp.* $\Delta_3(\infty)$ *the space* $\Delta_1(l)\backslash\{\infty\}$ *resp.* $\Delta_1(l)\backslash\{\pi(\infty, l)\}$ *is homeomorphic to* $\mathbb{R}^t$ *and closed in* $\Delta_2(\infty)$ *resp.* $\Delta_4(\infty)$*,*
- *the topologies on* $\Delta_1(\infty)$ *and* $\Delta_3(\infty)$ *are equivalent to the topologies induced by the Hausdorff metric on the space of closed subsets of* $\Delta_2(\infty)$ *resp.* $\Delta_4(\infty)$*,*
- *that the maps* $\pi_{i,j}$ *and* $\lambda_{i,j}$*, with* $(i,j) \in \{(0,3), (2,3), (4,1), (4,3)\}$*, as well as* ψ *are continuous.*

Moreover, suppose that a topology on $\mathcal{P} = \{\infty\} \cup \Delta_2(\infty) \cup \Delta_4(\infty)$ *is defined as follows: The topology on* $\Delta_4(\infty)$ *is the given one. In particular,* $\Delta_4(\infty)$ *is open in* $\mathcal{P}$*. If* $p \in \Delta_2(\infty)$*, choose* $l \in \Delta_3(\infty)$ *such that* $\lambda(\infty, l) \neq \infty \vee p$*. As a basis of neighbourhoods of* p *we take the sets* $N_{C,V} = \{q \in \mathcal{P} \setminus C \mid \lambda(q, l) \in V\}$*, where* $C \subset \Delta_4(\infty)$ *is a compact set and* $V \subset \Delta_2(l) \setminus \Delta_1(\pi(\infty, l))$ *is a neighbourhood of* $\lambda(p, l)$ *in* $\Delta_2(l) \cong \mathbb{R}^{s+t}$*.*

Finally, suppose that the topology on $\mathcal{L}$ *is defined via the Hausdorff metric on the space of closed subsets of* $\mathcal{P}$*.*

Then $\mathfrak{GQ}$ *becomes a compact quadrangle with parameters* (s, t) *such that the given topologies on the spaces* $\Delta_i(\infty)$ *are respected.* ۞

True, the above theorem is rather technical — but the proof is even more so. One major difficulty is to show that the topology on $\mathcal{P}$ is well defined, i.e. independent of the particular choice of the line l. The other major problem is to show that the flag space is closed in the space $\mathcal{P} \times \mathcal{L}$.

Our goal is to verify that the Lie geometry of a finite-dimensional connected Laguerre plane satisfies the above requirements. For this reason we have to have a closer look at the topological properties of the different spaces involved. There are natural identifications of the space of extended parallel classes and the space of extended tangent pencils with the space of parallel classes and the space of tangent pencils respectively. So, it suffices to define topologies on the space of parallel classes and on the space of tangent pencils to obtain topologies on $\Delta_1(\infty_\mathfrak{L})$ and $\Delta_3(\infty_\mathfrak{L})$.

Continuity of the maps $\pi_{i,j}$ and $\lambda_{i,j}$ follows from continuity of the geometric operations of the Laguerre plane except for $(i,j) = (4,3)$. This is no surprise, since for this case even the existence of the maps π and λ needed additional topological requirements.

Let $\mathfrak{L} = (\mathcal{S}, \mathcal{K})$ denote a locally compact Laguerre plane with parameter t, where $0 < t < \infty$. Then $t \in \{1,2\}$ by Corollary 1.2. Let $\mathfrak{GQ} = \mathfrak{GQ}(\mathfrak{L})$ denote the quadrangle constructed from the Laguerre plane $\mathfrak{L}$. Some basic results on topological Laguerre planes (see e.g. [13]: 3.8 and [48]: §1.Satz 3) stated in the setting of generalised quadrangles read as follows:

(3.6) Lemma: *Suppose $\mathfrak{L} = (\mathcal{S}, \mathcal{K})$ is a locally compact Laguerre plane with parameter t. Then:*

a) $\Delta_1(\infty_{\mathfrak{L}}) \cong \mathbb{S}_t$,
b) $\Delta_4(\infty_{\mathfrak{L}}) \cong \mathbb{R}^{3t}$,
c) for $l \in \Delta_1(\infty_{\mathfrak{L}})$ the truncated point row $\Delta_1(l) \setminus \{\infty_{\mathfrak{L}}\}$ is homeomorphic to $\mathbb{R}^t$ and closed in $\Delta_2(\infty_{\mathfrak{L}})$,
d) for $l \in \Delta_3(\infty_{\mathfrak{L}})$ the truncated point row $\Delta_1(l) \setminus \{\pi(\infty_{\mathfrak{L}}, l)\}$ is homeomorphic to $\mathbb{R}^t$ and closed in $\Delta_4(\infty_{\mathfrak{L}})$.

Since parallel classes and tangent pencils are closed subsets of $\Delta_2(\infty_{\mathfrak{L}}) = \mathcal{S}$ and $\Delta_4(\infty_{\mathfrak{L}}) = \mathcal{K}$ respectively, we can equip the spaces $\mathcal{G}$ and $\mathcal{B}$ with the topology induced by the Hausdorff metric. This induces topologies on $\Delta_1(\infty_{\mathfrak{L}})$ and $\Delta_3(\infty_{\mathfrak{L}})$.

The following lemma states the topological properties of the space $\mathcal{G}$. The proofs are rather straightforward using results of [12] and [13] and are omitted.

(3.7) Lemma: *Suppose $\mathfrak{L} = (\mathcal{S}, \mathcal{K})$ is a locally compact Laguerre plane with parameter t. Then:*

a) the map $\Gamma : \mathcal{S} \to \mathcal{G} : p \mapsto \Gamma(p)$ is continuous,
b) the map Γ restricted to the point set of a circle is a homeomorphism,
c) the space $\mathcal{G}$ is homeomorphic to the t-sphere $\mathbb{S}_t$.

The topological properties of $\mathcal{B}$ are less obvious. Let $\mathcal{F}_{\mathfrak{L}}$ be equipped with the topology it inherits from $\mathcal{S} \times \mathcal{K}$.

(3.8) Lemma: *Suppose $\mathfrak{L} = (\mathcal{S}, \mathcal{K})$ is a locally compact Laguerre plane with parameter t. Then:*

a) the flag space $\mathcal{F}_{\mathfrak{L}}$ is closed in $\mathcal{S} \times \mathcal{K}$,

b) if (s_i, k_i), $i \in \mathbb{N}$, is a sequence in $\mathcal{F}_{\mathfrak{L}}$ such that the sequence k_i, $i \in \mathbb{N}$, converges to a circle k, then the set $\{s_i \mid i \in \mathbb{N}\}$ has accumulation points and every such accumulation point is incident with k,

c) the space $\mathcal{B}$ is closed in the space of non-empty closed subsets of $\mathcal{K}$,

d) the map $B : \mathcal{F}_{\mathfrak{L}} \to \mathcal{B} : (s,k) \mapsto B(s,k)$ is surjective, continuous and open.

Proof:

a) Let (s_i, k_i), $i \in \mathbb{N}$, be a sequence in $\mathcal{F}_{\mathfrak{L}}$ converging to $(s,k) \in \mathcal{S} \times \mathcal{K}$. We have $s_i = \gamma(s_i, k_i)$ and thus continuity of parallel projection implies $s = \lim_{i\to\infty} s_i = \lim_{i\to\infty} \gamma(s_i, k_i) = \gamma(s,k)$, hence $(s,k) = (\gamma(s,k), k) \in \mathcal{F}_{\mathfrak{L}}$.

b) By the above, every accumulation point of $\{s_i \mid i \in \mathbb{N}\}$ is incident with k. Let $r_i := \gamma(s_i, k)$. Considered as a subset of $\mathcal{S}$, the circle k is compact ([13]: 2.2). Therefore, we may assume that the sequence r_i, $i \in \mathbb{N}$, converges to a point r incident with k. Thus, $\lim_{i\to\infty} s_i = \lim_{i\to\infty} \gamma(r_i, k_i) = \gamma(r,k) = r$.

c) Let g_i, $i \in \mathbb{N}$, be a sequence in $\mathcal{B}$ converging to some non-empty closed set $X \subset \mathcal{K}$. We have to show that X is a touching pencil as well.

Since $\lim_{i\to\infty} g_i = X \neq \emptyset$, there is a sequence $k_i \in g_i$, $i \in \mathbb{N}$, of circles converging to a circle $k \in X$. Let $s_i := \sigma(g_i)$. We have $g_i = B(s_i, k_i)$. By the above, there exists an accumulation point s of the set $\{s_i \mid i \in \mathbb{N}\}$, and (s,k) is a flag. Let $r \in \mathcal{S}$ be a point neither parallel to any of the points s_i nor to s. For every $i \in \mathbb{N}$ the map $\kappa_i : \Gamma(r) \to g_i = B(s_i, k_i) : t \mapsto \beta(s_i, k_i, t)$ is a homeomorphism ([12]: 1.1.b). Continuity of the map β implies that $B(s,k) = \beta(\{s\} \times \{k\} \times \Gamma(r))$ is contained in X. Therefore, s is the only accumulation point of the set $\{s_i \mid i \in \mathbb{N}\}$, since, by the above, every such point is incident with every circle of $B(s,k)$.

If l is a circle of X, then there are circles $l_i \in g_i$, $i \in \mathbb{N}$, converging to l. We have $l_i = \beta(s_i, k_i, \gamma(r, l_i))$, thus $l = \lim_{i\to\infty} l_i = \lim_{i\to\infty} \beta(s_i, k_i, \gamma(r, l_i)) = \beta(s, k, \gamma(r, l)) \in B(s,k)$. Hence $X = B(s,k)$, and X is indeed a touching pencil.

d) Obviously B is surjective. If $r \in \mathcal{S}$ and $(s,k) \in \mathcal{F}_{\mathfrak{L}}$ with $s \notin \Gamma(r)$, then $B(s,k) = \beta(\{s\} \times \{k\} \times \Gamma(r))$. Thus, continuity of β implies continuity of B.

Suppose that U is an open subset of $\mathcal{F}_{\mathfrak{L}}$, that the flag (s,k) is an element of U and that g_i, $i \in \mathbb{N}$, is a sequence in $\mathcal{B}$ converging to $B(s,k)$. Let $s_i := \sigma(g_i)$. By construction of the topology there is a sequence of circles $k_i \in g_i$, $i \in \mathbb{N}$, converging to k. By the above we obtain $\lim_{i\to\infty} s_i = s$. We may assume $(s_i, k_i) \in U$ for every $i \in \mathbb{N}$ and thus $g_i = B(s_i, k_i) \in B(U)$. Hence the map B is open.

Viewed in the Lie geometry $\mathfrak{G}\mathfrak{Q}(\mathfrak{L})$, Part d of the above lemma reads as follows.

(3.9) Corollary: *Let $\mathfrak{G}\mathfrak{Q}(\mathfrak{L})$ be the Lie geometry of a locally compact Laguerre plane $\mathfrak{L}$ with parameter $0 < t < \infty$. Then the restriction*

$$\rho : \{(a, b) \in \Delta_2(\infty_{\mathfrak{L}}) \times \Delta_4(\infty_{\mathfrak{L}}) \mid a \sim b\} \to \Delta_3(\infty_{\mathfrak{L}}) : (a, b) \mapsto a \vee b$$

is continuous and open. ☙

To prove continuity of the restrictions $\pi_{i,j}$ and $\lambda_{i,j}$ introduced earlier we need some more preparations. In fact, we need these preparations only for the restrictions $\pi_{4,3}$ and $\lambda_{4,3}$.

(3.10) Lemma: *Suppose that A is homeomorphic to $\mathbb{S}_1$ or $\mathbb{S}_2$ and suppose that τ is an involution on A which leaves exactly two points f_1 and f_2 fixed. Furthermore, suppose that $\Sigma \subset A$ is a continuous image of $\mathbb{S}_1$ and invariant under τ. Then the two fixed points are not contained in a common connected component of $A \setminus \Sigma$. Moreover, if A is homeomorphic to $\mathbb{S}_1$, then Σ contains at least one fixed point.*

Proof: If A is homeomorphic to $\mathbb{S}_1$, then τ is topologically equivalent to the reflection of the unit circle in $\mathbb{R}^2$ at a line through the origin ([3] or [21]). Therefore, $A \setminus \{f_1, f_2\}$ consists of precisely two connected components which are interchanged by τ. The set Σ is connected. Thus, if Σ contains no fixed point, then Σ has to be contained in one of the connected components of $A \setminus \{f_1, f_2\}$, contradicting the τ-invariance of Σ.

If A is homeomorphic to $\mathbb{S}_2$, then τ is topologically equivalent to the rotation of the unit sphere in $\mathbb{R}^3$ through π about a polar axis ([3] or [21]). The involution τ induces a twofold covering $c : A \to \mathbb{S}_2$ branched at f_1 and f_2. Since Σ is invariant under τ, the connected components of $A \setminus \Sigma$ are permuted by τ. If f_1, f_2 are contained in a common connected component K of $A \setminus \Sigma$, then K is invariant under τ and there is a path $\mu : [0, 1] \to c(K)$ from $c(f_1)$ to $c(f_2)$. The set $c^{-1}(\mu([0, 1])) \subset K$ separates A into precisely two connected components which are interchanged by τ. Since Σ is connected, it is contained in one of these connected components, contradicting the τ-invariance of Σ. ☙

Rather than showing continuity of $\lambda_{4,3}$ directly, we now prove, using the above lemma, that a map related to $\lambda_{4,3}$ is continuous.

(3.11) Lemma: *Suppose $\mathfrak{L} = (\mathcal{S}, \mathcal{K})$ is a locally compact Laguerre plane with parameter t. Then in the generalised quadrangle $\mathfrak{GQ}(\mathfrak{L})$ the map*

$$\eta : \Delta_4(\infty_{\mathfrak{L}}) \times \Delta_3(\infty_{\mathfrak{L}}) \setminus (\mathcal{F}_{\mathfrak{GQ}} \cup \pi^{-1}(\Delta_2(\infty_{\mathfrak{L}}))) \to \Delta_2(\infty_{\mathfrak{L}}) : (p, l) \mapsto \pi(\infty_{\mathfrak{L}}, \lambda(p, l))$$

is continuous.

Proof: Suppose $(p, l) \in \Delta_4(\infty_{\mathfrak{L}}) \times \Delta_3(\infty_{\mathfrak{L}}) \setminus (\mathcal{F}_{\mathfrak{GQ}} \cup \pi^{-1}(\Delta_2(\infty_{\mathfrak{L}})))$. Then $p \in \Delta_4(\infty_{\mathfrak{L}}) = \mathcal{K}$ is a circle not incident with $x := \sigma(l)$. By Lemma 3.3 there is a unique circle $k = \pi(p, l)$ in l touching p. Thus $l = B(x, k)$. Let y denote the touching point of k and p, i.e., $y = \pi(\infty_{\mathfrak{L}}, \lambda(p, l))$. We have to show that for every neighbourhood U of y in $\mathcal{S}$ there is a neighbourhood V of p in $\mathcal{K}$ and a neighbourhood W of l in $\mathcal{B}$, such that for every circle $v \in V$ and every tangent pencil $w \in W$ the point $\sigma(w)$ does not lie on v and the unique circle in w touching v given by Lemma 3.3 touches v in a point of U, i.e., such that $\pi(\{\infty_{\mathfrak{L}}\} \times \lambda(V \times W)) \subset U$ holds.

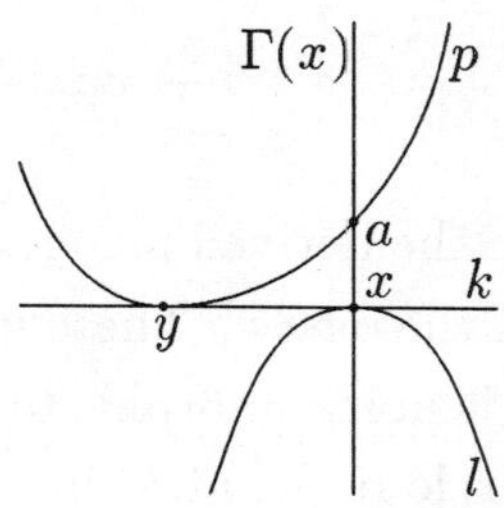

Fig. 3.1

To construct neighbourhoods V and W to a given U we take advantage of the local product structure of $\mathcal{S} = \Delta_2(\infty_{\mathfrak{L}})$ near y. Let $a := \Gamma(x) \cap p = \pi(p, x \vee \infty_{\mathfrak{L}})$ and consider $\mathcal{S} \setminus \Gamma(x) = B(a, p) \times \mathcal{G} \setminus \Gamma(x)$. Since $B(a, p) \cong \mathbb{R}^t \cong \mathcal{G} \setminus \Gamma(x)$ there are neighbourhoods A_1 of p in $B(a, p)$ and B_1 of $\Gamma(y)$ in $\mathcal{G} \setminus \Gamma(x)$ homeomorphic to the t-dimensional unit ball $\mathbb{D}_t$. Given any neighbourhood U of y in $\mathcal{S}$ we can choose A_1 and B_1 in such a way that the neighbourhood $U_1 := A_1 \times B_1$ of y in $\mathcal{S}$ is contained in U. The product structure of U_1 guarantees that for every $s \in p \cap U_1$ the intersection $\Gamma(s) \cap U_1$ is a neighbourhood of s in $\Gamma(s)$. Moreover, let G be a neighbourhood of p (regarded as a subset of $\mathcal{S}$) in $\mathcal{S}$ with $G \cap \Gamma(U_1) \subset U_1$. Choosing G small enough we can achieve that $k \cap G$ is contained in the interior of U_1. By continuity of β there are neighbourhoods $A_2 \subset A_1$ of p in $B(a, p)$ and $B_2 \subset B_1$ of $\Gamma(y)$ in $\mathcal{G} \setminus \Gamma(x)$ homeomorphic to $\mathbb{D}_t$, such that for every $s \in U_2 := A_2 \times B_2$ the circle $\beta(x, k, s)$ intersects G only in U_1. After possibly making G smaller we may assume that $G \cap \Gamma(U_2) \subset U_2$ and that $k \cap G$ is contained in the interior of U_2. Finally, by continuity of β, there is a neighbourhood U_y of y in $\Gamma(y)$ such that for $s \in U_y$ the circle $\beta(x, k, s)$ intersects the set G only in U_2.

Let $z \in k \setminus (\Gamma(x) \cup \Gamma(U_1))$. By continuity of the geometric operations there are neighbourhoods U_z of z in $\Gamma(z)$ and U_x of x in $\mathcal{S}$ with the following properties: The intersections $\Gamma(z) \cap \Gamma(U_x)$ and $\Gamma(y) \cap \Gamma(U_x)$ are empty, for every circle $h \in M := \alpha(U_x \times U_z \times U_y)$ and every point $c \in h \cap U_x$ we obtain $\beta(c, h, s) \cap G \in U_1$ if $s \in U_2$ and $\beta(c, h, s) \cap G \in U_2$ if $s \in U_y$.

The set M is a neighbourhood of k in $\mathcal{K}$ ([12]: 2.8.a). Let $V := \{v \in \mathcal{K} \mid v \subset G\}$ and $W := \{B(u,h) \mid h \in M, u \in U_x \cap h\}$. Then V is a neighbourhood of p in $\mathcal{K}$ ([13]: 1.4) and W is a neighbourhood of l in $\mathcal{B}$ (Lemma 3.10.d).

Suppose $v \in V$ is a circle and $w = B(u,h)$ with $h \in M$, $u \in U_x \cap h$ is an element of W. Let $d := v \cap \Gamma(y)$ and $h' := \beta(u,h,d)$. If v and h' have just the point d in common, then the point d is the required touching point which, by construction, is contained in $U_2 \subset U$. Otherwise $v \cap h' = \{d, e\}$ with $d \neq e$. By construction e lies in U_2. To find the circle in w touching v consider the map

$$\tau : v \to v : s \mapsto \begin{cases} (v \cap \beta(u,h,s)) \setminus \{s\} & \text{if } (v \cap \beta(u,h,s)) \setminus \{s\} \neq \emptyset, \\ s & \text{else.} \end{cases}$$

In the derived projective plane $\mathfrak{P}(\mathfrak{L}, u)$, this map corresponds to a bundle involution of the oval $\overline{v}$. These involutions are well known ([4]: 3.7). If $\mathfrak{L}$ has parameter 1, then v is homeomorphic to $\mathbb{S}_1$ and τ is topologically equivalent to the reflection of the unit circle in $\mathbb{R}^2$ at a line through the origin. If $\mathfrak{L}$ has parameter 2, then v is a 2-sphere and τ is topologically equivalent to the rotation through π about a polar axis. In both cases τ fixes precisely two points f_1, f_2. We may assume that f_1 corresponds to the point $\Gamma(u) \cap v$. Then f_2 corresponds to the required touching point.

Suppose $\Sigma \subset v$ is a continuous image of $\mathbb{S}_1$ invariant under τ. We can apply the results of Lemma 3.8. If v is a 1-sphere, then Σ has to contain at least one fixed point. If v is a 2-sphere, then Σ contains at least one fixed point, or the fixed points lie in different connected components of $v \setminus \Sigma$.

By construction we obtain $\tau(U_2 \cap v) \subset U_1 \cap v$. Since $U_2 \cap v$ is connected, there is a path $\mu : I \to U_2 \cap v$ joining d to e. The set $\Sigma := \mu(I) \cup \tau(\mu(I))$ is a continuous image of $\mathbb{S}_1$ contained in $U_1 \cap v$ and invariant under τ. The connected component of $v \setminus \Sigma$ containing f_1 is the unique connected component not contained in U_1. Thus, by the above considerations, the point f_2 is contained in $U_1 \subset U$. ☙

The above lemma enables us to prove continuity of the relevant restrictions of π and λ.

(3.12) Lemma: *Suppose $\mathfrak{L}$ is a locally compact Laguerre plane with parameter $0 < t < \infty$. Then the maps $\pi_{i,j}$ and $\lambda_{i,j}$, where $(i,j) \in \{(0,3),(2,3),(4,1),(4,3)\}$, as well as the map ψ are continuous.*

Proof: By construction of $\mathfrak{G}\mathfrak{Q}(\mathfrak{L})$, the map ψ corresponds to the map $\Gamma : \mathcal{S} \to \mathcal{G}$, which is continuous by Lemma 3.7.

The map $\pi_{0,3}$ corresponds to the map $\sigma : \mathcal{B} \to \mathcal{S}$, which is continuous since the map B is open (Lemma 3.10.d). Since $\lambda_{0,3} = \psi \circ \pi_{0,3}$, continuity of $\pi_{0,3}$ and ψ implies continuity of $\lambda_{0,3}$.

Next we prove that $\pi_{2,3}$ is continuous. If l_i, $i \in \mathbb{N}$, is a sequence in $\mathcal{B}$ converging to $l \in \mathcal{B}$, then, by Lemma 3.10.b and d, there is a sequence (s_i, k_i), $i \in \mathbb{N}$, in $\mathcal{F}_{\mathfrak{GQ}}$ converging to a flag (s, k) such that $l_i = B(s_i, k_i)$, $i \in \mathbb{N}$, and $l = B(s, k)$. Let p_i, $i \in \mathbb{N}$, be a sequence in $\mathcal{S}$ converging to a point $p \in \mathcal{S} \setminus \Gamma(s)$. Without loss of generality we may assume $p_i \notin \Gamma(s_i)$ for every $i \in \mathbb{N}$. Then $\pi_{2,3}(p_i, l_i) = \beta(s_i, k_i, p_i)$ for every $i \in \mathbb{N}$ and $\pi_{2,3}(p, l) = \beta(s, k, p)$. Thus, continuity of β implies continuity of $\pi_{2,3}$. Since $\lambda_{2,3}(p, l) = p \vee \pi_{2,3}(p, l) = \rho(p, \pi_{2,3}(p, l))$, continuity of $\lambda_{2,3}$ follows from continuity of $\pi_{2,3}$ and ρ.

We now investigate the map $\pi_{4,1}$. If l_i, $i \in \mathbb{N}$, is a sequence of parallel classes converging to a parallel class l, then there is a sequence of points $s_i \in l_i$, $i \in \mathbb{N}$, converging to a point $s \in l$. For every $i \in \mathbb{N}$ we obtain $l_i = \Gamma(s_i)$ and also $l = \Gamma(s)$. Let p_i, $i \in \mathbb{N}$, be a sequence of circles converging to a circle p. We have $\pi_{4,1}(p_i, l_i) = \gamma(s_i, p_i)$ for every $i \in \mathbb{N}$ and $\pi_{4,1}(p, l) = \gamma(s, p)$. Thus, continuity of γ implies continuity of $\pi_{4,1}$. Since $\lambda_{4,1} = \psi \circ \pi_{4,1}$, continuity of $\pi_{4,1}$ and ψ implies continuity of $\lambda_{4,1}$.

So far, we have only used continuity of the geometric operations of $\mathfrak{L}$ to show continuity of the maps in question. However, for the remaining case we have to employ the previous lemma. With η as in the previous lemma we obtain $\pi_{4,3}(p, l) = \pi_{2,3}(\eta(p, l), l)$ and $\lambda_{4,3}(p, l) = \lambda_{2,3}(\eta(p, l), l)$. Therefore, continuity of η, proved in Lemma 3.11, and continuity of $\pi_{2,3}$ and $\lambda_{2,3}$ proved above, implies continuity of the maps $\pi_{4,3}$ and $\lambda_{4,3}$. ⚙

We have now all requirements needed to employ Theorem 3.5.

(3.13) Theorem: *Suppose $\mathfrak{L} = (\mathcal{S}, \mathcal{K}, \mathcal{G})$ is a locally compact Laguerre plane with parameter $0 < s < \infty$. Then $s \in \{1, 2\}$ and*

$$\mathfrak{GQ}(\mathfrak{L}) = (\mathcal{S} \cup \mathcal{K} \cup \{\infty_{\mathfrak{L}}\}, \overline{\mathcal{B}} \cup \overline{\mathcal{G}}, \in)$$

equipped with topologies as constructed in Theorem 3.5 becomes a compact antiregular quadrangle with parameter s.

Proof: Except for antiregularity the result follows almost immediately from Theorem 3.5 if we identify $\infty_{\mathfrak{L}}$ with ∞. Then the sets $\Delta_i(\infty)$ are topological spaces and $\Delta_2(\infty) = \mathcal{S}$ as well as $\Delta_4(\infty) = \mathcal{K} \cong \mathbb{R}^{3s}$ are even metric spaces. Moreover,

$\Delta_1(\infty) \cong \mathbb{S}_s$ by Lemma 3.7.c. The required properties of the truncated line pencils hold by Lemma 3.6. Finally, the previous lemma shows that all maps mentioned in Theorem 3.5 are continuous. Thus, by Theorem 3.5, the generalised quadrangle $\mathfrak{GQ}(\mathfrak{L})$ equipped with topologies as described in Theorem 3.5 becomes a compact quadrangle with parameter s.

To verify antiregularity consider the space $\Delta_2(\infty_{\mathfrak{L}}) = \mathcal{S}$. If the Laguerre plane has parameter 1, then the point space $\mathcal{S}$ is homeomorphic to the cylinder $\mathbb{S}_1 \times \mathbb{R}$ ([13]), hence, by Lemma 2.3, the twisting number of $\mathfrak{GQ}(\mathfrak{L})$ is 0. If the Laguerre plane as parameter 2, then the point space $\mathcal{S}$ is homeomorphic to the tangent bundle of $\mathbb{S}_2$ ([48]), hence, by Lemma 2.3, the twisting number of $\mathfrak{GQ}(\mathfrak{L})$ is 2. Therefore, by Theorem 2.15, the quadrangle $\mathfrak{GQ}(\mathfrak{L})$ is antiregular. ☺

Theorem 3.1 states that for any point p of an antiregular generalised quadrangle $\mathfrak{GQ}$ the derived structure $\mathfrak{L}(\mathfrak{GQ}, p)$ is a Laguerre plane. Moreover, by Theorem 3.2 we know that the derived structure is even a Laguerre plane with parameter s if $\mathfrak{GQ}$ is a compact antiregular quadrangle with parameter s. A look at the construction involved reveals that it is inverse to the construction of the Lie geometry. More precisely, we have:

(3.14) Theorem: *If $\mathfrak{L}$ is a finite-dimensional connected Laguerre plane, then $\mathfrak{L}(\mathfrak{GQ}(\mathfrak{L}), \infty_{\mathfrak{L}}) = \mathfrak{L}$. If p is any point of an antiregular compact connected quadrangle $\mathfrak{GQ}$, then $\mathfrak{GQ}(\mathfrak{L}(\mathfrak{GQ}, p)) = \mathfrak{GQ}$ and $p = \infty_{\mathfrak{L}(\mathfrak{GQ},p)}$.*

If p, q are two points of a (compact) antiregular quadrangle, then $\mathfrak{L}(\mathfrak{GQ}, p)$ is (topologically) isomorphic to $\mathfrak{L}(\mathfrak{GQ}, q)$ if and only if there is a (continuous) automorphism of $\mathfrak{GQ}$ mapping p to q.

If p is a point of an antiregular (compact) quadrangle, then the group of (continuous) automorphisms of $\mathfrak{L}(\mathfrak{GQ}, p)$ consists precisely of those (continuous) automorphisms of $\mathfrak{GQ}$ that fix p.

Proof: The first part of the theorem follows immediately from the definitions of the constructions involved. Since incidence in the Lie geometry can be expressed entirely in terms of geometric operations of the Laguerre plane, the second part of the theorem is also obvious. ☺

Remark: The above theorem allows us to construct new Laguerre planes from a given Laguerre plane as follows. Take a Laguerre plane $\mathfrak{L}$ and a point different from $\infty_{\mathfrak{L}}$ in $\mathfrak{GQ}(\mathfrak{L})$. Then the derivation of $\mathfrak{GQ}(\mathfrak{L})$ at p yields again a Laguerre plane. We will come back to this line of thought in Chapter 6.

4 Möbius planes and antiregular quadrangles

Introduction

In this chapter we explore the connection between Möbius planes and generalised quadrangles. Unlike in the last chapter we start this time with the Möbius plane rather than with the quadrangle. The first guess that the Lie geometry of a Möbius plane is a generalised quadrangle turns out to be true in exceptional cases only. In particular, the Lie geometry of a flat Möbius plane is not a generalised quadrangle. The next idea is to consider oriented circles rather than circles. This leads to the "lifted Lie geometry". We will see that the lifted Lie geometry of a flat Möbius plane is indeed a generalised quadrangle.

There is also a rather natural way for the quadrangle constructed from a flat Möbius plane to become a topological, and indeed a compact quadrangle. This is much easier to verify than for Laguerre planes. We then proceed by characterising all compact quadrangles obtained from flat Möbius planes. Unlike in the case of Laguerre planes we do not obtain all compact antiregular quadrangles with parameter 1 in this fashion. We obtain precisely those compact antiregular quadrangles with parameter 1 which feature a certain involution. The involution is induced by reversing orientation of oriented circles.

This characterisation enables us to reverse the construction. Thus, given a compact antiregular quadrangle with such an involution we can construct a flat Möbius plane.

Some of the results of this chapter are not new. They appeared already in [11]. But the way we look at the situation and also the proofs are distinct from the treatment in [11]. In [11] Forst uses the fact that one can construct flat Laguerre planes from flat Möbius planes and the fact that one can construct compact quadrangles with parameter 1 from flat Laguerre planes to construct the quadrangle from a flat Möbius plane. We will proceed the other way round. We will construct the quadrangle directly from the Möbius plane. In Chapter 6 we then will use this construction and the constructions of the previous chapter to construct flat Laguerre planes from flat Möbius planes.

The reason why we proceed in this way is twofold. Firstly, it seems to be more natural and easier to understand. Secondly, it serves as a good background to understand the construction of a generalised quadrangle from a Minkowski plane.

4.1 The Lie geometry of a Möbius plane

We have seen in the previous chapter that the Lie geometry of a finite-dimensional connected Laguerre plane is a generalised quadrangle. So, our first attempt to construct a generalised quadrangle from a Möbius plane is to consider the Lie geometry of a Möbius plane. That is, we treat circles and points of a Möbius planes as one set of objects. There is also a natural relation on these objects. We will call this relation collinearity and it is defined as follows: two points are collinear if they coincide, a point is collinear with a circle if it is incident with the circle and two circles are collinear if they touch. Since we want to view the Lie geometry as a point-line geometry, we might define lines as maximal sets of pairwise collinear points or circles. It turns out that, at least for our purposes, this definition is not the right one. Instead, we define that extended touching pencils are the lines of the Lie geometry. More formally, we define the **Lie geometry $\mathfrak{Lie}(\mathfrak{Mo})$ of a Möbius plane $\mathfrak{Mo} = (\mathcal{S}, \mathcal{K})$** as follows:

$$\mathfrak{Lie}(\mathfrak{Mo}) := \left(\mathcal{S} \cup \mathcal{K}, \overline{\mathcal{B}}, \in\right).$$

The elements of $\mathcal{S} \cup \mathcal{K}$ are called points of the Lie geometry and the elements of $\overline{\mathcal{B}}$ are called lines of the Lie geometry.

Having defined the Lie geometry one now wonders what kind of geometry it is. Since, for example, two circles which intersect are not collinear, it is not some projective space. However, if l is an extended touching pencil and p a point of the Möbius plane different from the support s of l, then there is precisely one circle in l collinear with p, namely the circle $\beta(s, k, p)$, where k is any circle in l. Also, if l is an extended touching pencil and p a circle not contained in l but incident with the support s of l, then s is the unique element of l collinear with p. So one might conjecture that the Lie geometry of a Möbius plane is a generalised quadrangle. The only case we have not checked yet is the case where the point p of the Lie geometry is a circle of the Möbius plane not contained in the extended touching pencil l and not incident with the support of l. So, the Lie geometry of a Möbius plane is a generalised quadrangle if and only if for every tangent pencil l and any circle p not contained in l and not incident with the support of l there is precisely one circle in l touching p.

Whether or not this is true cannot be decided without some further information. There is a class of Möbius planes, namely Möbius planes connected to finite fields of characteristic two, where the Lie geometry is indeed a generalised quadrangle ([51]: Chap. 4). But this is a typical phenomenon of characteristic two. In the classical case there are precisely two circles in the extended touching pencil l touching the circle p. This situation is typical for finite Möbius planes of odd order but also for flat Möbius planes as will reveal the next lemma.

(4.1) Lemma: *Suppose p is a circle and l is a tangent pencil of a flat Möbius plane such that the support of l is not incident with p. Then precisely two circles in l touch p.*

Proof: Let $s := \sigma(l)$ denote the support of l and consider the derived projective plane $\mathfrak{P}(\mathfrak{Mo}, s)$. In the derived projective plane the touching pencil l together with the line at infinity l_∞ becomes a line pencil through some point t on l_∞. We have seen in Chapter 1 that l_∞ passes $\overline{p}$. Thus, by Theorem 1.1, there are precisely two lines through t tangent to $\overline{p}$. These two tangent lines correspond to two circles in l touching p. Conversely, any circle in l touching p corresponds to a line through t tangent to $\overline{p}$.

This result is the reason why we do not define lines of the Lie geometry to be maximal sets of pairwise collinear points of the Lie geometry. For there are two different types of such maximal sets, namely extended touching pencils and sets consisting of three mutually tangent circles not contained in a common touching pencil. Thus, defining lines to be maximal sets of pairwise collinear points one obtains a geometry with 'short' and 'long' lines. This is perhaps an interesting structure but not the subject of this book.

4.2 The lifted Lie geometry of a flat Möbius plane

The above result of the previous section implies that the Lie geometry of a flat Möbius plane is not a generalised quadrangle but a point-line geometry where for any non-incident point-line pair there are either precisely two points or precisely one point on l collinear with p. One can of course axiomatise this situation and study it. We will however not do so but instead try to get hold of the generalised quadrangle that seems to lurk behind this structure. The trick by which we can achieve this is the introduction of cycles or oriented circles.

So suppose $\mathfrak{Mo} = (\mathcal{S}, \mathcal{K})$ is a flat Möbius plane. Let $s \in \mathcal{S}$. Then $\mathfrak{A}(\mathfrak{Mo}, s)$ is an affine plane whose point space is homeomorphic to $\mathbb{R}^2$ by Corollary 1.2. The point space $\mathcal{S}$ is the one-point-compactification of the point space of the affine plane, where s is the point compactifying the space. Thus $\mathcal{S}$ is homeomorphic to the one-point-compactification of $\mathbb{R}^2$, hence to $\mathbb{S}_2$ (cf. [4]: 3.2). By Corollary 1.2, every circle is homeomorphic to $\mathbb{S}_1$. Therefore, by the Jordan curve theorem, the complement of every circle in $\mathcal{S}$ splits into two connected components, each of which is homeomorphic to $\mathbb{R}^2$. Moreover, if k_1 and k_2 are two circles of a flat Möbius plane touching each other, then for $i = 1, 2$ there is precisely one connected component A_i of $\mathcal{S} \setminus k_i$ such that $A_1 \subset A_2$. A **cycle** or **oriented circle** of a flat Möbius plane is a pair (k, A), where $k \in \mathcal{K}$ is a circle and A is one of the two connected components of $\mathcal{S} \setminus k$. We say that two **cycles (k, A) and (l, B) touch** if and only if k touches l and $A \subset B$ or $B \subset A$. Furthermore, we say that **a point $s \in \mathcal{S}$ is incident with a cycle (k, A)** if and only if s is incident with k in the Möbius plane. An **oriented touching pencil** or **cycle pencil** is a set of cycles such that any two cycles of this set touch and such that the set of corresponding circles is a touching pencil. Again, the **support $\sigma(b)$ of a cycle pencil** is the unique point incident with every cycle of the pencil and a cycle pencil together with its support is called an **extended cycle pencil**.

We denote the space of cycles by $\widetilde{\mathcal{K}}$, the space of cycle pencils and extended cycle pencils $\widetilde{\mathcal{B}}$ and $\widetilde{\widetilde{\mathcal{B}}}$ respectively. The flag space, i.e., the space of incident point-cycle pairs is denoted $\widetilde{\mathcal{F}}$. By **c** we denote the natural projection $c : \widetilde{\mathcal{K}} \to \mathcal{K} : (k, A) \mapsto k$ as well as the induced map $c : \widetilde{\mathcal{B}} \to \mathcal{B}$.

Suppose $\widetilde{k} = (k, A)$ is a cycle, s a point incident with $\widetilde{k}$ and t a point different from s. Let $l := \beta(s, k, t)$. Then there is precisely one component B of $\mathcal{S} \setminus l$ such that (l, B) touches (k, A). We denote (l, B) by $\boldsymbol{\widetilde{\beta}(s, \widetilde{k}, t)}$.

Let $\tau : \widetilde{\mathcal{K}} \to \widetilde{\mathcal{K}} : (k, A) \mapsto (k, \mathcal{S} \setminus (A \cup k))$ denote the map that reverses orientation. The following properties of the map τ are obvious.

(4.2) Lemma: *Let $\tau : \widetilde{\mathcal{K}} \to \widetilde{\mathcal{K}}$ as above. Then:*

a) the map τ is an involution,

b) two cycles c_1 and c_2 touch if and only if $\tau(c_1)$ and $\tau(c_2)$ touch. ۞

We now look at the analogy of Lemma 4.1.

(4.3) Lemma: *Suppose p is a cycle and l is a cycle pencil of a flat Möbius plane such that the support of l is not incident with p. Then precisely one cycle in l touches p.*

Proof: Let $k := c(p)$ and $b := c(l)$. By Lemma 4.1 there are precisely two circles x_1 and x_2 in b touching k. Let $y_1 := (x_1, A_1)$ and $y_2 := (x_2, A_2)$ denote the two cycles in l such that $c(y_i) = x_i$. Moreover, let B denote the connected component of $\mathcal{S} \setminus k$ such that $p = (k, B)$. After possibly reversing orientation we may assume $\sigma(l) \notin B$ and after possibly interchanging x_1 and x_2 we may assume $A_1 \subset A_2$. But then A_2 contains also $x_1 \setminus x_2$, since $x_1 \setminus x_2$ is connected and contained in the boundary of A_1. Thus A_2 contains also $k \cap x_1$, since by assumption $k \cap x_1 \neq x_1 \cap x_2 = \sigma(l)$. Hence A_2 contains $k \setminus x_2$. By our choice of B we have $B \cap x_2 = \emptyset$. Hence B is contained in one of the connected components of $\mathcal{S} \setminus x_2$, namely in the one containing $k \setminus x_2$, i.e., in A_2. Thus y_2 touches p. The same arguments show that B is contained in the connected component of $\mathcal{S} \setminus x_1$ containing $x_2 \setminus x_1$. Hence $B \cap A_1 = \emptyset$ and therefore y_1 does not touch p. Thus y_2 is the unique cycle in l touching p. ♋

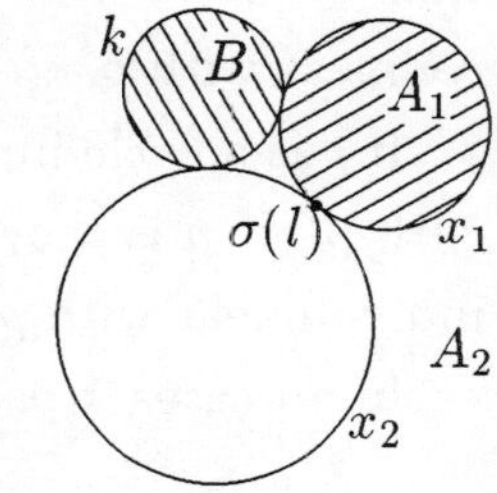

Fig. 4.1

This result suggests that instead of the Lie geometry one should consider the incidence geometry whose points are points and cycles of the Möbius plane and whose lines are extended cycle pencils. For reasons which will become apparent later, we will call this geometry the **lifted Lie geometry** of a flat Möbius plane, denoted $\widetilde{\mathfrak{Lie}}(\mathfrak{Mo})$. More precisely:

$$\widetilde{\mathfrak{Lie}}(\mathfrak{Mo}) := \left(\mathcal{P} := \mathcal{S} \cup \widetilde{\mathcal{K}}, \mathcal{L} := \widetilde{\widetilde{\mathcal{B}}}, \in\right)$$

where $\mathfrak{Mo}$ is a flat Möbius plane. This construction works only for flat Möbius planes, since flatness was essential for the definition of cycles. Over suitable fields one can define cycles also algebraically, but for an arbitrary Möbius plane there need not exist a lifted Lie geometry. The Lie geometry, however, can be defined for every Möbius plane.

The next result states that we succeeded in capturing the generalised quadrangle.

(4.4) Theorem: *The lifted Lie geometry $\widetilde{\mathfrak{Lie}}(\mathfrak{Mo})$ of a flat Möbius plane is a generalised quadrangle.*

Proof: We have to show that for every pair $(p, l) \in \mathcal{P} \times \mathcal{L}$ with $p \notin l$ there exists precisely one point $\pi(p, l)$ in l collinear with p and one line $\lambda(p, l)$ through p intersecting l. The line l is a cycle pencil whose support is denoted s. The point p can be a point or a cycle of the Möbius plane.

If $p \in \mathcal{S}$, then the condition that p is not incident with l says that p is different from s. Let k denote a cycle in l. Then $\pi(p, l) = \widetilde{\beta}(s, k, p)$ is the unique point on l collinear with p.

If p is a cycle incident with s, then $\pi(p, l) = s$ is the unique element of l collinear with p. If p is a cycle not incident with s, then there is precisely one cycle $\pi(p, l)$ in l collinear with p by Lemma 4.2.

In all cases $\lambda(p, l)$ is the extended cycle pencil spanned by p and $\pi(p, l)$. ⚙

Remark: The above proof shows some resemblance with the considerations in the last chapter. There also the topological requirements were only needed for the situation where p is a circle and l an extended touching pencil whose support is not incident with p.

Notation: Now that we know that the lifted Lie geometry of a flat Möbius plane is a generalised quadrangle, we denote the lifted Lie geometry of a flat Möbius plane $\mathfrak{Mo}$ by $\mathfrak{GQ}(\mathfrak{Mo})$.

4.3 Constructing topologies on the lifted Lie geometry

Naturally, a geometric structure obtained from some topological geometry should again be a topological geometry. So we want the quadrangle obtained from a flat Möbius plane to be a topological quadrangle, where the topologies on point- and line space should in some natural way be connected to the topologies on point- and circle space of the Möbius plane.

The set of points of the quadrangle obtained from a Möbius plane is the union of the set of points of the Möbius plane and the set of cycles. The point set comes already with a topology and the topology on the circle space defines a natural topology on the cycle space. However, we cannot simply say that the point space is the disjoint union of these two spaces, since then it would be disconnected. The classical example of a Möbius plane, where points can be viewed as circles with vanishing radius, suggests that the cycle space should be embedded as an open set in the point space of the lifted Lie geometry and that the point space of the Möbius space should be embedded in such a way, that the point space of the lifted Lie geometry becomes compact.

Circles of a flat Möbius plane are homeomorphic to $\mathbb{S}_1$, thus compact and therefore closed in $\mathcal{S}$. Hence, instead of the given topology on $\mathcal{K}$ we can also define a topology via the Hausdorff metric. Luckily, this yields again the given topology.

(4.5) Lemma: *If $\mathfrak{Mo}$ is a flat Möbius plane, then the topology on the circle space $\mathcal{K}$ induced by the Hausdorff metric on the space of closed non-empty subsets of $\mathcal{S}$ is equal to the given topology on $\mathcal{K}$.*

Proof: Suppose $k_i \in \mathcal{K}$, $i \in \mathbb{N}$, is a sequence of circles converging to some circle $k \in \mathcal{K}$ with respect to the given topology on $\mathcal{K}$. If $s \in k$, let l denote a circle through s not touching k. Since the set of pairs of properly intersecting circles is open ([26]: 4.3), infinitely many of the circles k_i intersect l. Thus, without loss of generality, we may assume that l intersects all of the circles k_i. By continuity of intersection this implies that the sets $l \cap k_i$ converge to $l \cap k$, which in turn induces that there are points $s_i \in k_i \cap l$ with $\lim_{i\to\infty} s_i = s$. On the other hand, if $s_i \in k_i$ is a sequence converging to some s, let $x \neq y$ denote two points on k different from s. By the above we know that there are points $x_i, y_i \in k_i$ converging to x and y respectively. Since the points s, x, y are mutually distinct, we also may assume that the point s_i, x_i, y_i are mutually distinct for every $i \in \mathbb{N}$. By continuity of the map α we get $k = \lim_{i\to\infty} k_i = \lim_{i\to\infty} \alpha(s_i, x_i, y_i) = \alpha(s, x, y)$ and thus $s \in k$. Therefore, the sequence k_i converges to k also with respect to the topology induced by the Hausdorff metric.

Conversely, suppose $k_i \in \mathcal{K}$, $i \in \mathbb{N}$, is a sequence of circles converging to some circle k with respect to the topology induced by the Hausdorff metric. Let $s, t, u \in k$ denote three points. Then there is a sequence of triples $(s_i, t_i, u_i) \in k_i^3$ converging to (s, t, u). We may assume $s_i \neq t_i \neq u_i \neq s_i$. Thus $k_i = \alpha(s_i, t_i, u_i)$ and $k = \alpha(s, t, u)$. By continuity of the map α the circles k_i converge to k also with respect to the given topology on $\mathcal{K}$. ♋

This enables us to define a natural topology on the point space of the Lie geometry of a flat Möbius plane. Namely, we identify every point $s \in \mathcal{S}$ with the set $\{s\} \subset \mathcal{S}$ and equip the space $\mathcal{S} \cup \mathcal{K}$ with the topology induced by the Hausdorff metric on the set of closed subsets of $\mathcal{S}$. Thus, by the preceding lemma and the fact that with respect to the Hausdorff metric any space X is homeomorphic to $\{\{x\} \mid x \in X\}$, we have homeomorphic copies of $\mathcal{S}$ and $\mathcal{K}$ in the point space of the Lie geometry.

Our next step is to prove that this construction yields a compact point space.

(4.6) Lemma: *The point space $\mathcal{S} \cup \mathcal{K}$ of the Lie geometry of a flat Möbius plane $\mathfrak{Mo} = (\mathcal{S}, \mathcal{K})$ equipped with the induced topology is compact.*

Proof: Suppose $k_i \in \mathcal{K}$ is a sequence of circles. Let $s_i \in k_i$. Since $\mathcal{S}$ is compact we may assume, after possibly considering only a subsequence, that there is some s with $s = \lim_{i \to \infty} s_i$. Next choose $t_i \in k_i$. Again we may assume that there is some $t \in \mathcal{S}$ such that $t = \lim_{i \to \infty} t_i$. If, for every choice of the sequence t_i, we get $t = s$, then $\lim_{i \to \infty} k_i = \{s\} \subset \mathcal{S} \cup \mathcal{K}$ with respect to the Hausdorff metric. Otherwise there exists some $t \neq s \in \mathcal{S}$ and some $t_i \in k_i$ with $\lim_{i \to \infty} t_i = t$. Let l denote a circle such that s and t are contained in different connected components of $\mathcal{S} \setminus l$. We may assume that the points t_i are in the same connected component of $\mathcal{S} \setminus l$ as t and that the points s_i are in the same connected component of $\mathcal{S} \setminus l$ as s. Then there are points $x_i \neq y_i$ such that $l \cap k_i = \{x_i, y_i\}$. We may assume that there is some $x \in \mathcal{S}$ with $\lim_{i \to \infty} x_i = x$. Then $\lim_{i \to \infty} k_i = \lim_{i \to \infty} \alpha(s_i, t_i, x_i) = \alpha(s, t, x) \in \mathcal{K}$. Thus every sequence of circles has a subsequence that converges either to a circle or to a point of the Möbius plane.

In the beginning of Section 4.2 we deduced from Corollary 1.2 that the point set of a flat Möbius plane is homeomorphic to $\mathbb{S}_2$. Hence it is compact and thus, with the above, the point space of the Lie geometry is also compact. ☉

Our main concern however is the lifted Lie geometry rather than the Lie geometry of a flat Möbius plane. First we have a look at the space of cycles. For a point $s \in \mathcal{S}$ let ${}_s\mathcal{K} := \{k \in \mathcal{K} \mid s \notin k\}$ denote the set of circles not incident with s. Let ${}_s\widetilde{\mathcal{K}} := c^{-1}({}_s\mathcal{K})$ denote the set of cycles not incident with s. For a circle $k \in {}_s\mathcal{K}$ the space $\mathcal{S} \setminus k$ splits into two connected components, one of which contains s and one of which does not contain s. Let A_k^+ denote the connected component of $\mathcal{S} \setminus k$ containing s and let A_k^- denote the connected component of $\mathcal{S} \setminus k$ not containing s. Let ${}_s\widetilde{\mathcal{K}}^+ := \{(k, A_k^+) \mid k \in {}_s\mathcal{K}\}$ and ${}_s\widetilde{\mathcal{K}}^- := \{(k, A_k^-) \mid k \in {}_s\mathcal{K}\}$. Then ${}_s\widetilde{\mathcal{K}}$ is the disjoint union of ${}_s\widetilde{\mathcal{K}}^+$ and ${}_s\widetilde{\mathcal{K}}^-$. Moreover, c maps ${}_s\widetilde{\mathcal{K}}^+$ as well as ${}_s\widetilde{\mathcal{K}}^-$ bijectively onto ${}_s\mathcal{K}$. We define topologies on ${}_s\widetilde{\mathcal{K}}^+$ and ${}_s\widetilde{\mathcal{K}}^-$ by requiring that the restrictions $c : {}_s\widetilde{\mathcal{K}}^+ \to {}_s\mathcal{K}$ and $c : {}_s\widetilde{\mathcal{K}}^- \to {}_s\mathcal{K}$ are homeomorphisms.

A topology on the whole of $\widetilde{\mathcal{K}}$ is defined as follows. Let s_i, $i \in \{1, 2, 3, 4\}$, denote four points not incident with a common circle. The spaces ${}_{s_i}\widetilde{\mathcal{K}}$ are topologised as described above. It is readily checked that on intersections ${}_{s_i}\widetilde{\mathcal{K}} \cap {}_{s_j}\widetilde{\mathcal{K}}$ the topologies coincide. Since, by choice of the points s_i, every circle in $\mathcal{K}$ misses at least one of the points s_i, we obtain a topology on $\widetilde{\mathcal{K}}$. The construction also yields that the map $c : \widetilde{\mathcal{K}} \to \mathcal{K}$ is a twofold covering.

Next we extend the map $c : \widetilde{\mathcal{K}} \to \mathcal{K}$ by the identity on $\mathcal{S}$ to a map $c : \mathcal{S} \cup \widetilde{\mathcal{K}} \to \mathcal{S} \cup \mathcal{K}$. Requiring this map to be continuous we obtain a topology on $\mathcal{P} = \mathcal{S} \cup \widetilde{\mathcal{K}}$.

We have to verify that with this topology the space $\mathcal{P}$ is metrisable. Choose $s_i \in \mathcal{S}$ as above. There are disjoint open neighbourhoods U_i of s_i such that no circle intersects all U_i. Let $\mathcal{K}_i := \{k \in \mathcal{K} \mid k \cap U_i = \emptyset\} = \bigcap_{s \in U_i} {}_s\mathcal{K}$ denote the set of circles missing U_i. Let $\widetilde{\mathcal{K}}_i^+ := \bigcap_{s \in U_i} {}_s\widetilde{\mathcal{K}}^+$ and $\widetilde{\mathcal{K}}_i^- := \bigcap_{s \in U_i} {}_s\widetilde{\mathcal{K}}^-$. Then $\mathcal{S} \setminus U_i \cup \mathcal{K}_i$ is a closed and therefore compact subset of $\mathcal{S} \cup \mathcal{K}$. The spaces $\mathcal{S} \setminus U_i \cup \widetilde{\mathcal{K}}_i^+$ and $\mathcal{S} \setminus U_i \cup \widetilde{\mathcal{K}}_i^-$ are homeomorphic to $\mathcal{S} \setminus U_i \cup \mathcal{K}$, thus also compact and closed in $\mathcal{P}$. Moreover, since $\mathcal{S} \setminus U_i \cup \mathcal{K}_i$ is metrisable, the spaces $\mathcal{S} \setminus U_i \cup \widetilde{\mathcal{K}}_i^+$ and $\mathcal{S} \setminus U_i \cup \widetilde{\mathcal{K}}_i^-$ are also metrisable. By [9]: 4.4.19 this implies that $\mathcal{P} = \bigcup_{i=1}^4((\mathcal{S} \setminus U_i \cup \widetilde{\mathcal{K}}_i^+) \cup (\mathcal{S} \setminus U_i \cup \widetilde{\mathcal{K}}_i^-))$ is metrisable.

In terms of sequences the topology can be described as follows. A sequence of cycles (k_i, A_i) converges to a cycle (k, A) if and only if $\lim_{i\to\infty}(A_i \cup k_i) = A \cup k$ with respect to the Hausdorff metric. A sequence of cycles (k_i, A_i) converges to a point $s \in \mathcal{S}$ if and only if $\lim_{i\to\infty} k_i = s$ with respect to the Hausdorff metric. This however does not imply that the topology on the point space of the lifted Lie geometry is induced by the Hausdorff metric, since if (k_i, A_i) converges to some point s and also $\lim_{i\to\infty}(A_i \cup k_i) = \{s\}$, then for the sequence $\tau(k_i, A_i) = (k_i, B_i)$ we obtain $\lim_{i\to\infty}(B_i \cup k_i) = \mathcal{S}$.

To verify that the lifted map $\widetilde{\beta}$ is continuous consider $(s, k, t) \in \mathcal{S} \times \mathcal{K} \times \mathcal{S}$ such that $s \in k$ and $t \neq s$. Let v denote a point that is neither incident with k nor with $\beta(s, k, t)$. Then there are neighbourhoods U_s of s and U_t of t in $\mathcal{S}$ and there is a neighbourhood U_k of k in $\mathcal{K}$ such that no circle in U_k is incident with u and such that for every triple $(x, y, z) \in U_s \times U_k \times U_t$ with $x \in y$ and $x \neq z$ the circle $\beta(x, y, z)$ is not incident with u. Let $V := \{(x, y, z) \in U_s \times U_k \times U_t \mid z \neq x \in y\}$. Let $W \subset V$ denote the connected component of V containing (s, k, t). Thus, with notation as above, U_k as well as $\beta(W)$ are connected subsets of ${}_u\mathcal{K}$. Let $c_+ : {}_u\widetilde{\mathcal{K}}^+ \to {}_u\mathcal{K}$ denote the restriction of c to ${}_u\widetilde{\mathcal{K}}^+$ and let $c_- : {}_u\widetilde{\mathcal{K}}^- \to {}_u\mathcal{K}$ denote the restriction of c to ${}_u\widetilde{\mathcal{K}}^-$. These restrictions are homeomorphisms. Let $W^+ := c_+^{-1}(W)$ and $W^- := c_-^{-1}(W)$. Then, for $(x, y, z) \in W^+$ we have $\widetilde{\beta}(x, y, z) = c_+^{-1}(\beta(x, c_+(y), z))$ and likewise, for $(x, y, z) \in W^-$ we have $\widetilde{\beta}(x, y, z) = c_-^{-1}(\beta(x, c_-(y), z))$. Thus $\widetilde{\beta}$ is continuous at $(s, \widetilde{k}, t)$ for every $\widetilde{k} \in c^{-1}(k)$. Since the triple (s, k, t) was chosen arbitrarily in the domain of definition of β, the map $\widetilde{\beta}$ is also continuous.

As a corollary to Lemma 4.6 and by the above considerations we obtain:

(4.7) Corollary: *The point space $\mathcal{P} = \mathcal{S} \cup \widetilde{\mathcal{K}}$ of the lifted Lie geometry of a flat*

Möbius plane $\mathfrak{Mo} = (\mathcal{S}, \mathcal{K})$ *is compact and the lifted map* $\widetilde{\beta}$ *is continuous.* ⚽

The next step is to define a topology on the line space of the lifted Lie geometry. Again, we want to use the Hausdorff metric, thus we have to verify that extended cycle pencils are closed in the point space

(4.8) Lemma: *Lines of the lifted Lie geometry of a flat Möbius plane are closed in the point space of the lifted Lie geometry and lines of the Lie geometry of a flat Möbius plane are closed in the point space of the Lie geometry.*

Proof: Let l denote an extended cycle pencil, let $s := \sigma(l)$ and suppose $k_i \in l$, $i \in \mathbb{N}$, is a sequence of cycles converging to some $k \in \mathcal{P}$. We may assume $k_i \neq s$ for every $i \in \mathbb{N}$. If $k \in \mathcal{S}$, then $k = s$, since $s \in c(k_i)$ for every $i \in \mathbb{N}$, thus $s \in \lim_{i\to\infty} k_i$.

If $k \in \widetilde{\mathcal{K}}$, choose $t \in c(k)$ different from s. By Lemma 4.5 there is a sequence of points $t_i \in c(k_i)$, $i \in \mathbb{N}$, converging to t. We may assume $t_i \neq s$, thus $k_i = \widetilde{\beta}(s, k_0, t_i)$. Therefore, by continuity of $\widetilde{\beta}$, we obtain $k = \lim_{i\to\infty} k_i = \lim_{i\to\infty} \widetilde{\beta}(s, k_0, t_i) = \widetilde{\beta}(s, k_0, t) \in l$.

This shows that extended cycle pencils, i.e., lines of the lifted Lie geometry, are closed. An analogous argument shows that extended touching pencils, i.e., lines of the Lie geometry are closed. ⚽

The above lemma enables us to equip the line space of the Lie geometry as well as the line space of the lifted Lie geometry with the topology induced by the Hausdorff metric on the space of closed subsets of the relevant point set. Almost immediately we obtain the following result.

(4.9) Lemma: *The line space* $\mathcal{L}$ *of the lifted Lie geometry and the line space of the Lie geometry of a flat Möbius plane are compact.*

Proof: We will only prove that the line space of the lifted Lie geometry is compact. The result for the line space of the Lie geometry can be shown analogously.

Suppose that $l_i \in \mathcal{L}$ is a sequence of lines. Let $s_i := \sigma(l_i)$. By compactness we may assume that there is some $s \in \mathcal{S}$ such that $\lim_{i\to\infty} s_i = s$. Choose $t \in \mathcal{S}$ different from s and let k_i denote the unique cycle in l_i incident with t. By compactness of $\mathcal{P}$ we may assume that there is some $k \in \mathcal{P}$ such that $\lim_{i\to\infty} k_i = k$. Since $s \neq t \in k$ we know $k \in \widetilde{\mathcal{K}}$. Let l denote the extended cycle pencil spanned by s and k, i.e., let $l := \{\widetilde{\beta}(s, k, x) \mid x \in \mathcal{S} \setminus \{s\}\} \cup \{s\}$. We claim $l = \lim_{i\to\infty} l_i$.

If $g \in l$ take $x \in g$ different from $s = \sigma(l)$. Then $g_i := \widetilde{\beta}(s_i, k_i, x) \in l_i$ and $\lim_{i\to\infty} g_i = \lim_{i\to\infty} \widetilde{\beta}(s_i, k_i, x) = \widetilde{\beta}(s, k, x) = g$. Conversely, if $g_i \in l_i$ is a sequence of cycles converging to some cycle g, then either $g = s \in l$ or there is some $x \in g$ different from s. In the latter case there exists a sequence of points $x_i \in g_i$ converging to x and thus $g = \lim_{i\to\infty} g_i = \lim_{i\to\infty} \widetilde{\beta}(s_i, k_i, x_i) = \widetilde{\beta}(s, k, x) \in l$.

The next result shows that extended touching pencils are homeomorphic to $\mathbb{S}_1$.

(4.10) Lemma: *Lines of the lifted Lie geometry as well as lines of the Lie geometry of a flat Möbius plane are homeomorphic to* $\mathbb{S}_1$.

Proof: Again we will only consider lines in the lifted Lie geometry. For lines in the Lie geometry the results are obtained analogously.

Suppose l is a line of the lifted Lie geometry of a flat Möbius plane. Let $s := \sigma(l)$, let $k \in l \setminus \{s\}$ and choose a circle z incident with s such that z and $c(k)$ do not touch i.e., such that z and $c(k)$ intersect properly. Then the map

$$\mu : z \setminus \{s\} \to l \setminus \{s\} : x \mapsto \widetilde{\beta}(s, k, x)$$

whose inverse map is

$$\mu^{-1} : l \setminus \{s\} \to z \setminus \{s\} : y \mapsto c(y) \cap k \setminus \{s\}$$

is a homeomorphism. Thus $l \setminus \{s\}$ is homeomorphic to a circle minus a point, hence it is homeomorphic to $\mathbb{R}$. By Lemma 4.8 and Corollary 4.7 the line l is closed in the compact point space, hence also compact. This implies that l is homeomorphic to the one-point-compactification of $\mathbb{R}$, i.e., to $\mathbb{S}_1$.

We are now in a position to prove that the generalised quadrangle obtained as the lifted Lie geometry of a flat Möbius plane is a topological quadrangle. This theorem appears already in [11], however there the proof is completely different from the one presented here.

(4.11) Theorem: *Suppose* $\mathfrak{Mo}$ *is a flat Möbius plane, i.e., a locally compact Möbius plane with parameter 1. Then the lifted Lie geometry* $\mathfrak{G}\Omega(\mathfrak{Mo})$ *of* $\mathfrak{Mo}$ *is a compact quadrangle with parameter 1.*

Proof: By Theorem 4.4 the lifted Lie geometry of a flat Möbius plane is a generalised quadrangle. By Corollary 4.7 and Lemma 4.9 the point space and the line space of the generlised quadrangle are compact. Furthermore, the space $\mathcal{F} \in \mathcal{P} \times \mathcal{L}$ of incident point-line pairs is closed in $\mathcal{P} \times \mathcal{L}$, since we have defined the topology on $\mathcal{L}$ via the Hausdorff metric on $\mathcal{P}$. This implies that if $(p_i, l_i) \in \mathcal{F}$, $i \in \mathbb{N}$, is a sequence converging to some $(p, l) \in \mathcal{P} \times \mathcal{L}$, then $p \in l$ by the definition of the topology on $\mathcal{L}$, thus $(p, l) \in \mathcal{F}$. Let $\mathcal{D} := \mathcal{P} \times \mathcal{L} \setminus \mathcal{F}$ denote the domain of definition of the map $\lambda \times \pi$. Then, by the above, the graph $\{(q, g, l, z) \in \mathcal{D} \times \mathcal{L} \times \mathcal{P} \mid \{(q, l), (z, l), (z, q)\} \subset \mathcal{F}\}$ is closed in $\mathcal{D} \times \mathcal{L} \times \mathcal{P}$. Therefore, $\lambda \times \pi$ is continuous ([8]: XI.2.7), thus the maps λ and π are also continuous.

By Lemma 4.10 lines of the quadrangle obtained are homeomorphic to $\mathbb{S}_1$. By Corollary 1.2 circles of a flat Möbius plane are homeomorphic to $\mathbb{S}_1$. Now suppose $k \in \widetilde{\mathcal{K}}$ is a point of the obtained quadrangle. Let $K \subset \mathcal{S}$ denote the points of the Möbius plane incident with $c(k)$. Then $K = k^{\perp} \cap \mathcal{S}$ and K is homeomorphic to $\mathbb{S}_1$. The same argument applies for $\tau(k)$, hence $K = k^{\perp} \cap \tau(k)^{\perp}$. The space $k^{\perp} \cap \tau(k)^{\perp}$ however is naturally homeomorphic to the space of lines incident with k ([45]: 2.2). Therefore, the line pencil through k is homeomorphic to $\mathbb{S}_1$ and hence the same holds for any other line pencil ([11]: 2.4). ♋

Remark: One also can define maps π and λ on the Lie geometry itself. For every non-incident point-line pair (p, l), we say that $\pi(p, l)$ consists of all points on l incident with p and that $\lambda(p, l)$ consists of all lines through p meeting l. We have seen that $\pi(p, l)$ and $\lambda(p, l)$ either contain both one element or contain both two elements. Then the Hausdorff metric defines a topology on the space of sets consisting of one or two points and the space of sets consisting of one or two lines. One sees readily that with these topologies the maps π and λ are continuous. So we can say that the Lie geometry of a flat Möbius plane is a topological incidence geometry.

4.4 Characterising quadrangles obtained from flat Möbius planes

We have seen in Chapter 2 that compact quadrangles with parameter 1 are, up to duality, antiregular. By Lemma 2.17 we know that a compact quadrangle with parameter 1 is antiregular if and only if there are three (necessarily pairwise non-collinear) points s, t and u such that there is no point collinear with s, t and u

Now let k denote a cycle and s and t two points of a flat Möbius plane such that s and t are contained in different connected components of $\mathcal{S} \setminus c(k)$. Every circle through s touching $c(k)$ has empty intersection with the connected component of $\mathcal{S} \setminus c(k)$ containing t. In particular, there is no circle through s and t touching $c(k)$, hence there is no point of the lifted Lie geometry collinear with s, t and k. So we obtain the following result:

(4.12) Theorem: *The lifted Lie geometry $\mathfrak{GQ}(\mathfrak{Mo})$ of a flat Möbius plane $\mathfrak{Mo}$ is an antiregular compact quadrangle with parameter 1.* ☺

Notation: Suppose $\mathfrak{Mo}$ is a flat Möbius plane. Reversing orientation of cycles defines an involution on $\widetilde{\mathcal{K}}$. Extending this involution by the identity on $\mathcal{S}$ to $\widetilde{\mathcal{K}} \cup \mathcal{S}$, we obtain an involution on $\mathcal{P}$ denoted $\tau_{\mathfrak{Mo}}$.

(4.13) Lemma: *The quadrangle $\mathfrak{GQ}(\mathfrak{Mo})$ obtained as the lifted Lie geometry of a flat Möbius plane $\mathfrak{Mo} = (\mathcal{S}, \mathcal{K})$ is an antiregular compact quadrangle with parameter 1 allowing an involutory collineation $\tau_{\mathfrak{Mo}}$ whose fixed line set is empty and whose fixed point set Σ is homeomorphic to $\mathbb{S}_2$, such that $s^\perp \cap t^\perp \cap u^\perp \neq \emptyset$ for any three points $s, t, u \in \Sigma$ and such that every line contains precisely one fixed point.*

Proof: Suppose $\mathfrak{Mo} = (\mathcal{S}, \mathcal{K})$ is a flat Möbius plane. Then the involution $\tau := \tau_{\mathfrak{Mo}}$ has the desired properties. The fixed point set of τ is $\mathcal{S} \subset \mathcal{P}$, hence homeomorphic to $\mathbb{S}_2$. If $s, t, u \in \mathcal{S}$ are three points, then $s^\perp \cap t^\perp \cap u^\perp$ consists of the two cycles k and $\tau(k)$ where $c(k) = c(\tau(k)) = \alpha(s, t, u)$. Moreover, no line of the quadrangle, i.e., no extended cycle pencil is fixed, since no cycle is collinear with its image $\tau(k)$. However, every line contains precisely one fixed point, namely the support of the corresponding extended cycle pencil. By Theorem 4.12 the quadrangle obtained is antiregular and compact with parameter 1. ☺

The above lemma suggests that all antiregular compact quadrangles with an involution as above are connected to Möbius planes. In turns out that this is true and that we even can neglect the topological assumptions.

(4.14) Theorem: *Suppose $\mathfrak{GQ} = (\mathcal{P}, \mathcal{L})$ is a thick antiregular quadrangle with an involutory collineation τ such that no line is fixed by τ, such that every line contains a point fixed by τ and such that for any three points s, t and u in the fixed point set Σ of τ the centre $s^\perp \cap t^\perp \cap u^\perp$ is not empty. Then the derived structure*

$$\mathfrak{Mo}(\mathfrak{GQ}, \tau) := \left(\Sigma, \{\Sigma \cap k^\perp \mid k \in \mathcal{P} \setminus \Sigma\}, \in\right) = \left(\Sigma, \{k^\perp \cap \tau(k)^\perp \mid k \in \mathcal{P} \setminus \Sigma\}, \in\right)$$

is a Möbius plane.

Proof: No two points in Σ are collinear, since any two collinear fixed points would span a line fixed by τ.

Suppose $k \in \mathcal{P}$ is a point not fixed by τ. Let l denote a line though k. Then there is precisely one point x on l fixed by τ. However, since x is collinear with k, the point $\tau(k)$ is collinear with $\tau(x) = x$, hence $l \cap \Sigma = \{x\} = \{\pi(\tau(k), l)\}$. We can do this for every line through k and therefore obtain $k^\perp \cap \Sigma = k^\perp \cap \tau(k)^\perp$ and, for reasons cf symmetry, also $\tau(k)^\perp \cap \Sigma = k^\perp \cap \tau(k)^\perp$.

If s, t and u are three points in Σ, then $s^\perp \cap t^\perp \cap u^\perp \neq \emptyset$ by assumption, hence, by antiregularity, there are two points x and y such that $s^\perp \cap t^\perp \cap u^\perp = \{x, y\}$. Neither x nor y are in Σ, since no two points in Σ are collinear. However, since s, t and u are fixed by τ, the set $s^\perp \cap t^\perp \cap u^\perp = \{x, y\}$ is invariant under τ, hence $y = \tau(x)$. This implies that $\alpha(s, t, u) = x^\perp \cap \tau(x)^\perp$ with $\{x, \tau(x)\} = s^\perp \cap t^\perp \cap u^\perp$ is the unique circle through s, t and u.

If $s, t \in \Sigma$ and $k \in \mathcal{P} \setminus \Sigma$ such that $t \in k^\perp \cap \Sigma$, then $k^\perp \cap \Sigma = \tau(k)^\perp \cap \Sigma$ but $l^\perp \cap \Sigma \neq k^\perp \cap \Sigma$ for every $l \in \mathcal{P} \setminus (\Sigma \cup \{k, \tau(k)\})$ by the above. Let $m := \pi(s, t \vee k)$. Then $m \notin \Sigma$ and $\{s, t\} \in m^\perp \cap \Sigma$. Moreover, if $s \not\sim k$, then $m^\perp \cap k^\perp \cap \Sigma = \Sigma \cap (m \vee k)$ contains just one point, namely t. Thus $m^\perp \cap \Sigma$ touches $k^\perp \cap \Sigma$. If $s \sim k$, then $m = k$ and $m^\perp \cap \Sigma$ touches $k^\perp \cap \Sigma$ trivially. Moreover, $\pi(s, t \vee \tau(k)) = \tau(\pi(s, t \vee k))$, hence $\beta(t, (k^\perp \cap \Sigma), s) = \beta(t, (\tau(k)^\perp \cap \Sigma), s) = \pi(s, t \vee k)^\perp \cap \Sigma = \tau(\pi(s, t \vee k))^\perp \cap \Sigma = \pi(s, t \vee \tau(k))^\perp \cap \Sigma$ is the unique circle through s touching $k^\perp \cap \Sigma$ in t.

Thickness of the quadrangle implies that the derived geometry is not degenerate. Therefore, the derived structure is indeed a Möbius plane. ☺

Remark: Note that any thick antiregular quadrangle with an involution as above yields a Möbius plane, but not every Möbius plane gives rise to a generalised quadrangle. The construction of the quadrangle requires additional properties of the Möbius plane. In our case these are topological properties. For Laguerre planes the situation is similar. There, every point of a thick antiregular quadrangle gives rise to a Laguerre plane, but only Laguerre planes with additional properties give rise to quadrangles.

However, if we restrict our attention to finite-dimensional connected geometries, the situation is much nicer. More precisely we have the following theorem:

(4.15) Theorem: *Suppose* $\mathfrak{Mo} = (\mathcal{S}, \mathcal{K})$ *is a flat Möbius plane, i.e., a locally compact Möbius plane with parameter 1. Then the quadrangle* $\mathfrak{GQ}(\mathfrak{Mo})$ *obtained as the lifted Lie geometry of* $\mathfrak{Mo}$ *is an antiregular compact quadrangle with parameter 1 allowing an involutory collineation* $\tau_{\mathfrak{Mo}}$ *whose fixed line set is empty and whose fixed* *point set* Σ *is homeomorphic to* $\mathbb{S}_2$, *such that* $s^\perp \cap t^\perp \cap u^\perp \neq \emptyset$ *for any three points* $s, t, u \in \Sigma$ *and such that every line contains precisely one fixed point.*

Conversely, if $\mathfrak{G}\mathfrak{Q} = (\mathcal{P}, \mathcal{L})$ *is an antiregular compact quadrangle with parameter 1 and an involution* τ *as above, then the derived structure*

$$\mathfrak{Mo}(\mathfrak{G}\mathfrak{Q}, \tau) = (\Sigma, \{\Sigma \cap k^\perp \mid k \in \mathcal{P} \setminus \Sigma\}, \in) = (\Sigma, \{k^\perp \cap \tau(k)^\perp \mid k \in \mathcal{P} \setminus \Sigma\}, \in)$$

where Σ *denotes the fixed point set of* τ, *is a flat Möbius plane.*

The two constructions are inverse to each other, namely

$$\mathfrak{G}\mathfrak{Q}(\mathfrak{Mo}(\mathfrak{G}\mathfrak{Q}, \tau)) = \mathfrak{G}\mathfrak{Q}$$
$$\tau_{\mathfrak{Mo}(\mathfrak{G}\mathfrak{Q}, \tau)} = \tau$$

for any quadrangle $\mathfrak{G}\mathfrak{Q}$ *and involution* τ *as above and*

$$\mathfrak{Mo}(\mathfrak{G}\mathfrak{Q}(\mathfrak{Mo}), \tau_{\mathfrak{Mo}}) = \mathfrak{Mo}$$

for every flat Möbius plane $\mathfrak{Mo}$.

Proof: The first part of the theorem follows from Lemma 4.13. By Theorem 4.14 we know that the derived structure $\mathfrak{Mo}(\mathfrak{G}\mathfrak{Q}, \tau)$ is a Möbius plane. For $k \in \mathcal{P} \setminus \Sigma$ the space $k^\perp \cap \tau(k)^\perp$ is homeomorphic to the line pencil through k ([45]: 2.2), hence to $\mathbb{S}_1$. Thus every circle of the derived structure is homeomorphic to $\mathbb{S}_1$. The point space Σ of the Möbius plane inherits a topology from $\mathcal{P}$ and circles are homeomorphic to $\mathbb{S}_1$, thus, by [55]: 8.1, the derived structure is a flat Möbius plane, where the topology on the circle space is induced by the Hausdorff metric. One might also verify directly that the Möbius plane is flat using the fact that in compact antiregular quadrangles the map that assigns to every triad its centre is continuous (Proposition 2.4).

The remaining statements follow directly from the constructions. ∎

Given an antiregular compact quadrangle with parameter 1 and some involution τ it may be rather tedious to check whether or not τ meets the above requirements. There is however much redundancy in this condition and it can be replaced by several equivalent ones.

(4.16) Proposition: *Suppose* τ *is an involutory collineation on a compact antiregular quadrangle with parameter 1 leaving no line fixed. Then the following conditions are equivalent:*

a) The derived structure $\mathfrak{Mo}(\mathfrak{GQ},\tau)$ *is a flat Möbius plane.*

b) The fixed point set Σ *of* τ *is homeomorphic to* $\mathbb{S}_2$*, every line contains precisely one fixed point and* $s^\perp \cap t^\perp \cap u^\perp \neq \emptyset$ *for any three points* $s, t, u \in \Sigma$.

c) The fixed point set of τ *is homeomorphic to* $\mathbb{S}_2$.

d) There are three fixed points s, t *and* u *such that* $s^\perp \cap t^\perp \cap u^\perp \neq \emptyset$.

e) Every line contains a point fixed by τ.

Proof: The equivalence of a) and b) was shown in the previous theorem. The implications b) $\to$ c) and b) $\to$ d) as well as b) $\to$ e) are trivial.

Assume every line contains a fixed point. Since no line is fixed, every line contains precisely one fixed point. Take a point z not fixed by τ and let s, t and u denote the fixed points on three different lines through z. Then $z \in s^\perp \cap t^\perp \cap u^\perp$. This proves the implication e) $\to$ d).

Assume that the fixed point set Σ of τ is homeomorphic to $\mathbb{S}_2$. Suppose $s \in \Sigma$ is a point fixed by τ and let l denote a line through s. Then $l' := l \setminus \{s\} \cong \mathbb{R}$ and $\Sigma' := \Sigma \setminus \{s\} \cong \mathbb{R}^2$. The map $\mu : \Sigma' \to l' : x \mapsto \pi(t, l)$ is continuous, hence it cannot be injective. Take $t \neq u \in \Sigma'$ such that $\mu(t) = \mu(u)$, then $\mu(t) \in s^\perp \cap t^\perp \cap u^\perp$. This proves the implication c) $\to$ d).

Finally, we prove the implication d) $\to$ a) which then implies the claimed result. This proof however requires some more arguments.

Let Σ denote the fixed point set of τ and suppose that $s, t, u \in \Sigma$ are three points with $s^\perp \cap t^\perp \cap u^\perp \neq \emptyset$.

For the further discussion we need to know how τ acts on $s^\perp$. Since the quadrangle is antiregular and compact with parameter 1, the space $s^\perp \setminus \{s\}$ is homeomorphic to $\mathbb{R} \times \mathbb{S}_1$ (Proposition 2.4 and Lemma 2.3) and for every point z not collinear with s the space $s^\perp \setminus (z^\perp \cup \{s\})$ consists of two connected components. Moreover, if z and y are two points such that $s^\perp \cap y^\perp \cap z^\perp$ contains precisely two points, then $(s^\perp \cap y^\perp) \setminus z^\perp$ consists of two connected components which are contained in different connected components of $s^\perp \setminus (z^\perp \cup \{s\})$. For $a, b \in \Sigma$ the involution τ acts without fixed points on the space $a^\perp \cap b^\perp$ which is homeomorphic to $\mathbb{S}_1$, thus it acts equivalently to the rotation of $\mathbb{S}_1$ through 180°.

So now consider the given points s, t and u. The two components of $(s^\perp \cap u^\perp) \setminus t^\perp$ get interchanged by τ, since otherwise there would be fixed points in $s^\perp \cap u^\perp$. But then τ also interchanges the two components of $s^\perp \setminus (t^\perp \cup \{s\})$. Now suppose $z \notin s^\perp$ is a point such that $s^\perp \cap t^\perp \cap z^\perp = \emptyset$. Then $s^\perp \cap z^\perp$ is contained entirely in one connected component of $s^\perp \setminus (t^\perp \cup \{s\})$ and $s^\perp \cap \tau(z)^\perp = \tau(s^\perp \cap z^\perp)$ is contained entirely in the other connected component of $s^\perp \setminus (t^\perp \cup \{s\})$. Therefore $\tau(z)^\perp \cap s^\perp \neq z^\perp \cap s^\perp$

which implies $z \neq \tau(z)$. Hence, for every point z fixed by τ different from s and t the intersection $s^\perp \cap t^\perp \cap z^\perp$ is not empty. Since no lines are fixed, no two fixed points are collinear. By antiregularity, the intersection $s^\perp \cap t^\perp \cap z^\perp$ has therefore to contain precisely two points which get interchanged by τ. Replacing t by z in the above argument we obtain that for any two points y and z in Σ different from s the intersection $s^\perp \cap z^\perp \cap y^\perp$ contains two points. Finally, replacing s by z and z by y in the above argument we obtain that for any three points $x, y, z \in \Sigma$ the intersection $z^\perp \cap y^\perp \cap x^\perp$ contains precisely two points k and $\tau(k)$.

To see that there are points in Σ different from s, t and u consider the two points k and $\tau(k)$ in $s^\perp \cap t^\perp \cap u^\perp$. The set $k^\perp \cap \tau(k)^\perp$ is homeomorphic to the line pencil through k, hence to $\mathbb{S}_1$. Furthermore the set $k^\perp \cap \tau(k)^\perp$ is invariant under τ. Thus τ acts as an involution on $k^\perp \cap \tau(k)^\perp$ with at least three fixed points, namely s, t and u. Therefore τ fixes $k^\perp \cap \tau(k)^\perp$ pointwise, since every involution on $\mathbb{S}_1$ having three fixed points is the identity, as can be deduced from the intermediate value theorem. Thus $k^\perp \cap \tau(k)^\perp \subset \Sigma$. Since no line is fixed, every line contains at most one fixed point. If l is a line through k, then $l \cap \Sigma = \pi(\tau(k), l)) \in k^\perp \cap \tau(k)^\perp$. This implies $k^\perp \cap \Sigma = k^\perp \cap \tau(k)^\perp = \tau(k)^\perp \cap \Sigma$.

Our next claim is that for any point $k \in s^\perp \cap t^\perp$ the set $k^\perp \cap \tau(k)^\perp$ is contained in Σ. So suppose $k^\perp \cap \tau(k)^\perp \not\subset \Sigma$. Again, τ acts on $k^\perp \cap \tau(k)^\perp$ with fixed points s and t. But these are the only fixed points, since otherwise τ would induce the identity on $k^\perp \cap \tau(k)^\perp$, i.e., $k^\perp \cap \tau(k)^\perp \subset \Sigma$, contradicting the assumption. However, an involution on a space homeomorphic to $\mathbb{S}_1$ having precisely two fixed points is equivalent to the reflection of $\mathbb{S}_1$ at some line through the origin. Now let l denote a line through s neither incident with k nor with $\tau(k)$. Then the map

$$\mu : k^\perp \cap \tau(k)^\perp \to l : z \mapsto \begin{cases} \pi(z, l) & z \neq s \\ s & z = s \end{cases}$$

is a homeomorphism ([43] 2.1) and the induced involution $\sigma := \mu \circ \tau \circ \mu^{-1}$ on l fixes precisely the points s and $\pi(t, l)$ and interchanges the two connected components of $l \setminus \{s, \pi(t, l)\}$. We have seen above, that the space $s^\perp \setminus (t^\perp \cup \{s\})$ consists of two connected components A and B. Furthermore, $s^\perp \setminus (t^\perp \cup \{s\} \cup (s \vee k) \cup (s \vee \tau(k)))$ consists of four connected components A_1, A_2, B_1 and B_2 such that $A_i \in A$ and $B_i \in B$ and such that $s \vee A_i = s \vee B_i$. Now let $z \in k^\perp \cap \tau(k)^\perp$. Then $s^\perp \cap z^\perp \setminus t^\perp$ consists of two components C and D one of which lies in A while the other lies in B. Also one of the connected components of $s^\perp \cap z^\perp \setminus t^\perp$ lies entirely in $A_1 \cup B_1$, while the other lies in $A_2 \cup B_2$. So we may assume $C \in A_1$ and $D \in B_2$. We have seen above that τ acts as a rotation on $s^\perp \cap t^\perp$, so it interchanges $A_1 \cup B_1$ and

$A_2 \cup B_2$. Furthermore, τ interchanges A and B, thus $\tau(C) \in B_2$ and $\tau(D) \in A_1$, in particular $\tau(z)^\perp \cap B_1 = \emptyset$. On the other hand, if l is a line through s as above such that $l \cap A_1 \neq \emptyset$, then $\mu(z) \in A_1$ and $\sigma(\mu(z)) = \mu(\tau(z)) = \tau(z)^\perp \cap l \in B_1$ hence $\tau(z)^\perp \cap B_1 \neq \emptyset$, contradicting the above. So, all in all, we find that τ cannot act on $k^\perp \cap \tau(k)^\perp$ leaving only s and t fixed, hence τ fixes $k^\perp \cap \tau(k)^\perp$ pointwise.

Together with the fact that for any $z \in \Sigma$ the intersection $s^\perp \cap t^\perp \cap z^\perp$ contains two points, we obtain $\Sigma = \bigcup_{k \in s^\perp \cap t^\perp}(k^\perp \cap \tau(k)^\perp)$. One could deduce from this that Σ is homeomorphic to $\mathbb{S}_2$. We will not do so, since we get this property of Σ later on for free. Our next step is to verify that every line contains a fixed point.

Suppose l is a line. Then there are fixed points s and t not incident with l. Let $x \in s^\perp \cap t^\perp$ such that $\{x, \tau(x)\} \cap \{\pi(s, \lambda(t,l)), \pi(t, \lambda(s,l))\} = \emptyset$. Consider the square projection $\mu : s^\perp \cap t^\perp \to l : v \mapsto \pi(v,l)$. The image $\mu(s^\perp \cap t^\perp)$ is an interval in l with endpoints $\pi(s,l)$ and $\pi(t,l)$. Let $a := \mu(x)$ and $b := \mu(\tau(x))$. By our choice of a and b we know that $\{\mu(a), \mu(b)\} \cap \{\pi(s,l), \pi(t,l)\} = \emptyset$. If $a = b$ then a is the point on l fixed by τ. If $a \neq b$ let A denote the closed interval of l with endpoints a and b contained in the image $\mu(s^\perp \cap t^\perp)$. Then $\mu^{-1}(A)$ consists of two disjoint connected components C and D and the restriction of μ to either C or D is a homeomorphism. Since τ acts as a rotation on $s^\perp \cap t^\perp$ the points x and $\tau(x)$ are in different connected components of $\mu^{-1}(A)$. So we may assume that there are points c and d such that x and c are the endpoints of C and such that $\tau(x)$ and d are the endpoints of D. We do not know whether $\tau(c) = d$, but we know that $\tau(C) \subset D$ or $D \subset \tau(C)$. After possibly interchanging x and $\tau(x)$ we may assume $\tau(C) \subset D$ hence also $B := \mu(\tau(C)) \subset A$. Let μ_C denote the restriction of μ to C and consider the map $\nu : A \to B : w \mapsto \mu(\tau({\mu_C}^{-1}(w)))$. By the Brouwer fixed-point theorem ([1]: 8.14) this map has a fixed point v. However, if v is a fixed point of ν, then $\mu^{-1}(v) = \{\mu_C(v), \tau(\mu_C(v))\}$, thus $v \in \{k^\perp \cap \tau(k)^\perp \mid k \in s^\perp \cap t^\perp\} = \Sigma$ is a point on l fixed by τ. By previous considerations we know that there is at most one fixed point on every line, hence we now know that every line is incident with precisely one fixed point.

Next suppose k is a point not fixed by τ. Then every line through k is incident with precisely one fixed point which has to be also collinear with $\tau(k)$. Hence $k^\perp \cap \Sigma = k^\perp \cap \tau(k)^\perp = \tau(k)^\perp \cap \Sigma$. In particular, this tells us that circles of the derived structure are homeomorphic to $\mathbb{S}_1$.

We are now in a position to readily check that the derived structure satisfies the axioms of a Möbius plane. The non-degeneracy conditions are fulfilled trivially. If x, y and z are three points in Σ, then we have seen above that there is a point k

not fixed by τ such that $x^\perp \cap y^\perp \cap z^\perp = \{k, \tau(k)\}$. Thus $\alpha(x, y, z) = k^\perp \cap \Sigma = k^\perp \cap \tau(k)^\perp = \tau(k)^\perp \cap \Sigma$ is the unique circle incident with x, y and z. Since the centre $x^\perp \cap y^\perp \cap z^\perp$ depends continuously on $\{x, y, z\}$ by Proposition 2.4 and since $k^\perp \cap \Sigma$ depends continuously on $k \notin \Sigma$ with respect to the Hausdorff metric, the map α is continuous. If $x, y \in \Sigma$ and $k \in \mathcal{P} \setminus \Sigma$ such that $y \in k^\perp \cap \Sigma$, then $\beta(y, k^\perp \cap \Sigma, x) = \pi(x, y \vee k)^\perp \cap \Sigma$ is the unique circle through x touching $k^\perp \cap \Sigma$ in y. Continuity of the operations of the quadrangle induces that the map β is also continuous.

Thus the derived structure is a topological Möbius plane where the point space is equipped with a topology such that circles are homeomorphic to $\mathbb{S}_1$. A result of Wölk ([55]: 8.1) says that then the Möbius plane is a flat Möbius plane, where the topology on the circle space is defined via the Hausdorff metric. ◈

Remark: In the above proof topological arguments are essential. Thus it provides no hint whether or not the requirements of Theorem 4.14 can also be weakened in the general case.

The automorphism group $\mathrm{Aut}(\mathfrak{Mo})$ of a flat Möbius plane $\mathfrak{Mo}$ and the automorphism group $\mathrm{Aut}(\mathfrak{GQ}(\mathfrak{Mo}))$ of the corresponding quadrangle are closely related. Standard arguments from group theory prove that an automorphism $\varphi \in \mathrm{Aut}(\mathfrak{GQ}(\mathfrak{Mo}))$ acts on $\mathfrak{Mo}$, i.e., leaves the fixed point set of $\tau_{\mathfrak{Mo}}$ invariant, if and only if φ commutes with $\tau_{\mathfrak{Mo}}$. Moreover, if τ is an involution of an antiregular compact quadrangle with parameter 1 and properties specified in Proposition 4.16 or an involution of a thick antiregular quadrangle with properties specified in Theorem 4.14, then this involution is already determined by its fixed point set Σ as follows. For any point q not in Σ there are at least three points $s, t, u \in \Sigma$ collinear with q. By antiregularity $s^\perp \cap t^\perp \cap u^\perp$ contains precisely one other point besides q. This other point is $\tau(q)$.

Therefore, denoting for an element α of a group G the subgroup of elements of G commuting with α by $\mathrm{Cs}_\alpha(G)$, we obtain the following results:

(4.17) Corollary: *Suppose $\mathfrak{GQ}$ is an antiregular compact quadrangle with parameter 1 and suppose τ is an involution acting on $\mathfrak{GQ}$ as specified in Proposition 4.16. Then*

$$\mathrm{Aut}(\mathfrak{Mo}(\mathfrak{GQ}, \tau)) = \mathrm{Cs}_\tau(\mathrm{Aut}(\mathfrak{GQ}))/\langle\tau\rangle.$$

Conversely, if $\mathfrak{Mo}$ is a flat Möbius plane, then

$$\mathrm{Aut}(\mathfrak{Mo}) = \mathrm{Cs}_{\tau_{\mathfrak{Mo}}}(\mathrm{Aut}(\mathfrak{GQ}(\mathfrak{Mo})))/\langle\tau\rangle.$$

◈

Again, standard arguments yield the following results on automorphisms not commuting with $\tau_{\mathfrak{Mo}}$.

(4.18) Corollary: *Suppose $\mathfrak{GQ}$ is an antiregular compact quadrangle with parameter 1 and suppose τ_1 is an involution acting on $\mathfrak{GQ}$ as specified in Proposition 4.16. Suppose τ_2 is another involution on $\mathfrak{GQ}$. Then τ_2 acts on $\mathfrak{GQ}$ as specified in Proposition 4.16 and $\mathfrak{Mo}(\mathfrak{GQ}, \tau_1) \cong \mathfrak{Mo}(\mathfrak{GQ}, \tau_2)$ if and only if there is an automorphism ψ of $\mathfrak{GQ}$ such that $\tau_2 = \psi\tau_1\psi^{-1}$.* ⚙

In the general, i.e., non-topological case we obtain:

(4.19) Corollary: *Suppose $\mathfrak{GQ}$ is a thick antiregular quadrangle. If τ is an involution acting on $\mathfrak{GQ}$ as specified in Theorem 4.14, then*

$$\mathrm{Aut}(\mathfrak{Mo}(\mathfrak{GQ}, \tau)) = \mathrm{Cs}_\tau(\mathrm{Aut}(\mathfrak{GQ}))/\langle\tau\rangle.$$

Suppose τ_2 is another involution on $\mathfrak{GQ}$. Then τ_2 acts on $\mathfrak{GQ}$ as specified in Theorem 4.14 and $\mathfrak{Mo}(\mathfrak{GQ}, \tau_1) \cong \mathfrak{Mo}(\mathfrak{GQ}, \tau_2)$ if and only if there is an automorphism ψ of $\mathfrak{GQ}$ such that $\tau_2 = \psi\tau_1\psi^{-1}$. ⚙

5 Minkowski planes and antiregular quadrangles

Introduction

In this chapter we will investigate the connection between Minkowski planes and generalised quadrangles. It turns out that the connection is similar as the connection between Möbius planes and generalised quadrangles. Namely, the lifted Lie geometry of a locally compact Minkowski plane with parameter 2 is a compact quadrangle with parameter 2. For flat Minkowski planes the situation is more complex. The lifted Lie geometry of a locally compact Minkowski plane with parameter 1 consists of *two* generalised quadrangles with parameter 1. The surprising fact that we obtain two generalised quadrangles from one flat Minkowski plane is due to the fact that the circle space of a flat Minkowski plane consists of two connected components.

The difficult part is to define cycles or oriented circles. Unlike in the case of flat Möbius planes there is no easy way to define an orientation for circles, not even for flat Minkowski planes. The key to success is to view orientation as a twofold covering of the circle space. This point of view allows us to define oriented circles for finite-dimensional connected Minkowski planes.

However, before we can introduce a twofold covering we first have to check that the spaces in question allow such a covering. So, great parts of this chapter are devoted to the investigation of the topological properties of point space and line space of finite-dimensional connected Minkowski planes. We then define the lifted Lie geometry of a finite-dimensional connected Minkowski plane and show that it is a generalised quadrangle if the parameter of the Minkowski plane is 2 and that it consists of two generalised quadrangles if the parameter of the Minkowski plane is 1. These quadrangles can be made topological quadrangles in a natural way. It turns out that if we start with a locally compact Minkowski plane with parameter 2 or 1 we obtain a compact antiregular quadrangle with parameter 2 or two compact antiregular quadrangles with parameter 1 respectively.

As in the case of Möbius planes we then characterise all quadrangles that arise via the lifted Lie geometry of finite-dimensional connected Minkowski planes. Again, we find that these are all compact antiregular quadrangles which allow involutions with certain properties. Conversely, given a compact antiregular quadrangle with such an involution one can construct a finite-dimensional Minkowski plane.

5.1 The point space and parallel classes

We first collect some known topological properties of finite-dimensional connected Minkowski planes. The classical model of locally compact Minkowski plane with parameter 1 or 2 is obtained as the geometry of non-trivial plane sections of a hyperboloid in the real or complex three-dimensional projective space respectively (see Section 1.1). At least as far as the point space is considered, the classical examples are typical as the following result from [49] shows.

(5.1) Proposition: *The point space of a finite-dimensional connected Minkowski plane* $\mathfrak{Mi} = (\mathcal{S}, \mathcal{K})$ *is two- or four-dimensional and compact. More precisely, circles are homeomorphic to the 1-sphere* $\mathbb{S}_1$ *or to the 2-sphere* $\mathbb{S}_2$ *and the point space is homeomorphic to* $\mathbb{S}_1 \times \mathbb{S}_1$ *or to* $\mathbb{S}_2 \times \mathbb{S}_2$ *respectively.* ⚙

From this result we can deduce the topology on parallel classes (cf. [35]: 1.2).

(5.2) Corollary: *Plus- and minus-parallel classes are homeomorphic to circles, hence to* $\mathbb{S}_t$ *if the Minkowski plane has topological parameter* t.

Proof: Let $s \in \mathcal{S}$ be a point and $k \in \mathcal{K}$ a circle. The map $\mu : \Gamma_+(s) \to k : x \mapsto \gamma_-(x, k)$ is continuous. To see that the inverse map exists and is continuous let $x \in k$ and choose a point $u \in \Gamma_+(s) \setminus \Gamma(x)$. Then the restriction $\mu^{-1} : k \setminus \Gamma(u) \to \Gamma_+(s) : v \mapsto \delta(u, v)$ is continuous. Since x was chosen arbitrarily, this proves the lemma. ⚙

We will use the above results frequently without referring to them.

5.2 The circle space

Next we investigate the space of circles. We will see that the circle space of a locally compact Minkowski plane with parameter 1 or 2 is homeomorphic to the circle space of the real or complex Minkowski plane respectively. However, our main purpose in the investigation of the circle space is to show that, in the topological case, there exists a twofold covering of the circle space such that the covering space has the same number of connected components as the circle space.

We first introduce coordinates for the circle space $\mathcal{K}$ of a Minkowski plane. Let s, t and u denote three mutually non-parallel points of a Minkowski plane and let

$$\boldsymbol{Tr}(\boldsymbol{\Gamma_-(s)}) := \{(a, b, c) \in \Gamma_-(s)^3 \mid a \neq b \neq c \neq a\}$$

denote the set of triples of mutually different points of $\Gamma_-(s)$. Then the map

$$\kappa : Tr(\Gamma_-(s)) \to \mathcal{K} : (a, b, c) \mapsto \alpha(a, \delta(b, t), \delta(c, u))$$

is a bijection. Its inverse is

$$\kappa^{-1} : \mathcal{K} \to Tr(\Gamma_-(s)) : k \mapsto (\gamma_-(s, k), \delta(\gamma_-(t, k), s), \delta(\gamma_-(u, k), s)).$$

If $\mathfrak{Mi}$ is a topological Minkowski plane, then κ is a homeomorphism, since all the maps involved to describe κ and κ^{-1} are continuous. So we can identify every circle with a triple of pairwise different points of $\Gamma_-(s)$ and vice versa. If $\mathfrak{Mi}$ is a locally compact Minkowski plane with parameter t, then every minus-parallel class is homeomorphic to $\mathbb{S}_t$, thus $\mathcal{K}$ is homeomorphic to the space of triples of mutually different points of $\mathbb{S}_t$. For $t = 1, 2$ these spaces are well-known: They are homeomorphic to $PGL(2, \mathbb{R})$ and $PSL(2, \mathbb{C})$ respectively, since these groups act sharply triply transitively on $\mathbb{S}_1$ and $\mathbb{S}_2$ respectively. The last claim can be verified by elementary methods if one considers $\mathbb{S}_1$ as the projective line over $\mathbb{R}$ and $\mathbb{S}_2$ as the projective line over $\mathbb{C}$.

The group $PGL(2, \mathbb{R})$ has two connected components which are homeomorphic to $PSL(2, \mathbb{R})$. A polar decomposition of $SL(2, \mathbb{R})$ into $SO(2, \mathbb{R})$ and triangular matrices shows that $SL(2, \mathbb{R})$, and hence also $PSL(2, \mathbb{R})$, is homeomorphic to $\mathbb{S}_1 \times \mathbb{R}^2$ ([20]). In particular, the fundamental group of every connected component of $PGL(2, \mathbb{R})$ is $\mathbb{Z}$. Thus for every $n \in \mathbb{N}$ there exist an (up to homeomorphisms) unique n-fold covering. If $n = 2$, then the map $P : SL(2, \mathbb{R}) \to PSL(2, \mathbb{R})$ induces a natural twofold covering. For our original problem this means that the circle space $\mathcal{K}$ of a flat Minkowski plane splits into two connected components $\mathcal{K}_1$ and $\mathcal{K}_2$ which allow twofold coverings. Geometrically, the homotopy equivalence from $\mathcal{K}_i$ to $\mathbb{S}_1$ can be described by projection onto the first coordinate.

The space $PSL(2, \mathbb{C})$, also known as the space of Möbius transformations, is connected and its unique connected twofold covering space is $SL(2, \mathbb{C})$ which is simply connected. Again, the covering map $P : SL(2, \mathbb{C}) \to PSL(2, \mathbb{C})$ induces a twofold covering of the circle space. Hence the circle space $\mathcal{K}$ of a locally compact Minkowski plane with parameter 2 is connected and there exists a unique connected twofold covering space $\widetilde{\mathcal{K}}$.

The following theorem summarises our results on the circle space of a topological Minkowski plane.

(5.3) Theorem: *The circle space $\mathcal{K}$ of a locally compact Minkowski plane with parameter 2 is homeomorphic to $PSL(2, \mathbb{C})$, thus it is connected and the unique twofold* *covering space $\widetilde{\mathcal{K}}$ is simply connected. The circle space $\mathcal{K}$ of a flat Minkowski plane is homeomorphic to $PGL(2,\mathbb{R})$. Thus it splits into two connected components $\mathcal{K}_1$ and $\mathcal{K}_2$ which allow unique twofold coverings with connected covering spaces.* ☺

Notation: We will denote the two connected components of the circle space $\mathcal{K}$ of a flat Minkowski plane by $\boldsymbol{\mathcal{K}_1}$ and $\boldsymbol{\mathcal{K}_2}$ and the twofold coverings by $c_i : \widetilde{\mathcal{K}}_i \to \mathcal{K}_i$, $i = 1, 2$. The twofold covering of the circle space of a locally compact Minkowski plane with parameter 1 or 2 is denoted $c : \widetilde{\mathcal{K}} \to \mathcal{K}$.

We have seen in Proposition 5.1 that the point space of a finite-dimensional connected Minkowski plane is metrisable and that circles are compact, hence closed set of points. Thus we can equip the spaces $\mathcal{K}$ also with the topology induced by the Hausdorff metric on the space of closed non-empty subsets of $\mathcal{S}$.

(5.4) Lemma: *If $\mathfrak{Mi}$ is a finite-dimensional connected Minkowski plane, then the topology on the circle space $\mathcal{K}$ induced by the Hausdorff metric on the space of closed non-empty subsets of $\mathcal{S}$ is equal to the given topology on $\mathcal{K}$.*

Proof: Suppose $k_i \in \mathcal{K}$, $i \in \mathbb{N}$, is a sequence of circles converging to some circle $k \in \mathcal{K}$ with respect to the given topology on $\mathcal{K}$. If $s \in k$, then $s_i := \gamma_+(s, k_i) \in k_i$, $i \in \mathbb{N}$, is a sequence of points converging to s by continuity of γ_+. On the other hand, if $s_i \in k_i$ is a sequence converging to some s, then $s_i = \gamma_+(s_i, k_i)$, hence $s = \gamma_-(s, k) \in k$ by continuity of γ_+. Therefore, the sequence k_i converges to k also with respect to the topology induced by the Hausdorff metric.

Conversely, suppose $k_i \in \mathcal{K}$, $i \in \mathbb{N}$, is a sequence of circles converging to some circle k with respect to the Hausdorff metric. Let $s, t, u \in k$ denote three points. Then there is a sequence of triples $(s_i, t_i, u_i) \in k_i^3$ converging to (s, t, u). We may assume $s_i \neq t_i \neq u_i \neq s_i$. Thus $k_i = \alpha(s_i, t_i, u_i)$ and $k = \alpha(s, t, u)$. Therefore, by continuity of the map α, the circles k_i converge to k also with respect to the given topology on $\mathcal{K}$. ☺

The following properties of certain subsets of the circle space of a locally compact Minkowski plane with parameter 1 or 2 follow readily from the coordinatisation of the circle space by $PGL(2,\mathbb{R})$ and $PSL(2,\mathbb{C})$ respectively.

(5.5) Lemma:

a) Touching pencils of a locally compact Minkowski plane with parameter t are homeomorphic to $\mathbb{R}^t$. In particular, they are connected and simply connected.

b) In a flat Minkowski plane the space $\mathcal{K}_s$ of circles through a point s and the space $\mathcal{K}_{s,t}$ of circles through two non-parallel points s and t split into two connected components and the different connected components of $\mathcal{K}_s$ or $\mathcal{K}_{s,t}$ lie in different connected components of $\mathcal{K}$. ֍

A question we will encounter later is, whether in locally compact Minkowski planes with parameter 2 the preimage of the set of circles through two given points under the twofold covering map c is connected or not. To answer this question consider the space

$$\Delta := \left\{ \begin{pmatrix} a & 0 \\ 0 & a^{-1} \end{pmatrix} \mid a \in \mathbb{C} \setminus \{0\} \right\} \subset SL(2, \mathbb{C})$$

of diagonal matrices of $SL(2,\mathbb{C})$. Then $P\Delta \subset PSL(2,\mathbb{C})$ is isomorphic to the space of Möbius transformations of the form $z \mapsto bz$ with $b \in \mathbb{C} \setminus \{0\}$, i.e., to the set of Möbius transformations fixing 0 and ∞. Thus $P\Delta$ can be identified with the set

$$\left\{ (s, \delta(t,s), b) \in \Gamma_-(s)^3 \mid b \in \Gamma_-(s) \setminus \{s, \delta(t,s)\} \right\} \subset Tr(\Gamma_-(s))$$

which in turn corresponds to the set of circles through s and t. Let $\mathcal{K}_{s,t}$ denote the set of circles incident with s and t, then by the above $\mathcal{K}_{s,t}$ and $\widetilde{\mathcal{K}}_{s,t} := c^{-1}(\mathcal{K})$ are homeomorphic to $\mathbb{C} \setminus \{0\}$. Since Δ contains the kernel of P, the restriction $c : \widetilde{\mathcal{K}}_{s,t} \to \mathcal{K}_{s,t}$ is again a twofold covering. The points s and t can be any non-parallel points of a locally compact Minkowski plane with parameter 2, thus we obtain the following result:

(5.6) Lemma: *If s and t are two non-parallel points of a locally compact Minkowski plane $\mathfrak{Mi}$ with parameter 2, then the restriction $c : \widetilde{\mathcal{K}}_{s,t} := c^{-1}(\mathcal{K}_{s,t}) \to \mathcal{K}_{s,t} := \{k \in \mathcal{K} \mid \{s,t\} \subset k\}$ is again a twofold covering and $\widetilde{\mathcal{K}}_{s,t}$ as well as $\mathcal{K}_{s,t}$ is homeomorphic to $\mathbb{C} \setminus \{0\}$.* ֍

Remark: It should be stressed that although the circle spaces of locally compact Minkowski planes with parameter 1 and of locally compact Minkowski planes with parameter 2 allow twofold coverings, these coverings are quite different in nature. For example, if s and t are two non-parallel points of a flat Minkowski plane, then every connected component of $\mathcal{K}_{s,t}$ is homeomorphic to $\mathbb{R}$, in particular simply connected. Therefore, the preimage of each connected component of $\mathcal{K}_{s,t}$ under the covering map splits into two connected components, hence the preimage of $\mathcal{K}_{s,t}$ consists of two disjoint copies of $\mathcal{K}_{s,t}$, quite in contrast to the results stated in Lemma 5.6.

5.3 The other spaces

Our next step is to define topologies on the flag space $\mathcal{F}$, on the spaces $\mathcal{G}_+$ and $\mathcal{G}_-$ of plus- and minus-parallel classes and on the space $\mathcal{B}$ of touching pencils. Furthermore, we will show that with these topologies the maps σ and B are continuous. Recall that σ maps a tangent pencil to its support and that B maps a flag (s,k) to the tangent pencil spanned by s and k.

The flag space $\mathcal{F}$ inherits its topology from $\mathcal{S}\times\mathcal{K}$. Since we want to define the topologies on $\mathcal{G}_+$, $\mathcal{G}_-$ and $\mathcal{B}$ using the Hausdorff metric, we first have to show that parallel classes and tangent pencils are closed (cf. [35]: 1.7 and 1.21).

(5.7) Lemma: *Suppose $\mathfrak{Mi}=(\mathcal{S},\mathcal{K})$ is a topological Minkowski plane. Then:*

a) Plus- and minus-parallel classes are closed in $\mathcal{S}$.
b) For any two non-parallel points $t,s\in\mathcal{S}$ there exist neighbourhoods N_s of s and N_t of t in $\mathcal{S}$ such that no point in N_s is parallel to any point in N_t.
c) The flag space $\mathcal{F}$ is closed in $\mathcal{S}\times\mathcal{K}$.
d) Touching pencils are closed in $\mathcal{K}$.

Proof: a) Let $g\in\mathcal{G}_+$. Take $s\in g$. By continuity of Γ_+ the set $\Gamma_+^{-1}(s)=g\times\mathcal{K}_s$ is closed in $\mathcal{S}\times\mathcal{K}$, thus g is closed in $\mathcal{S}$. The argument for minus-parallel classes is analogous.

b) Let $k\in\mathcal{K}$ denote a circle incident with s and t. There exist disjoint neighbourhoods M_s of s and M_t of t in k. By continuity of the maps $\gamma_{+,k}:\mathcal{S}\to k: s\mapsto\gamma_+(s,k)$ and $\gamma_{-,k}:\mathcal{S}\to k: s\mapsto\gamma_-(s,k)$, the sets $N_s:=\gamma_{+,k}^{-1}(M_s)\cap\gamma_{-,k}^{-1}(M_s)$ and $N_t:=\gamma_{+,k}^{-1}(M_t)\cap\gamma_{-,k}^{-1}(M_t)$ have the required properties.

c) Suppose $(s_i,k_i)\in\mathcal{F}$ is a sequence of flags converging to a pair $(s,k)\in\mathcal{S}\times\mathcal{K}$. Choose two non-parallel points $u,v\in\mathcal{K}\setminus\Gamma(s)$. We may assume that also $u,v\notin\Gamma_-(s_i)$, for every $i\in\mathbb{N}$. Let $u_i:=\gamma_-(u,k_i)$ and $v_i:=\gamma_-(v,k_i)$. Then $k=\lim_{i\to\infty}k_i=\lim_{i\to\infty}\alpha(s_i,u_i,v_i)=\alpha(\lim_{i\to\infty}s_i,\lim_{i\to\infty}u_i,\lim_{i\to\infty}v_i)=\alpha(s,u,v)$. In particular $s\in k$, thus $(s,k)\in\mathcal{F}$.

d) Let $b\in\mathcal{B}$. Take $k\in b$ and let $s:=\sigma(b)$. By continuity of β the set $\beta^{-1}(s,k)=\{s\}\times b\times\mathcal{S}\setminus\Gamma(s)$ is closed in $\mathcal{F}\times\mathcal{S}\setminus\Gamma(s)$. Hence $\{s\}\times b$ is closed in $\mathcal{F}$, thus also closed in $\mathcal{S}\times\mathcal{K}$ by the above, therefore b is closed in $\mathcal{K}$. ∎

We have seen in the previous section (Proposition 5.1, Theorem 5.3) that point space and circle space of a finite-dimensional connected Minkowski plane are metrisable. Thus we can equip the spaces $\mathcal{G}_+$ and $\mathcal{G}_-$ with the topology induced by the Hausdorff metric on the space of closed non-empty subsets of $\mathcal{S}$ and the spaces $\mathcal{B}$ with the topology induced by the Hausdorff metric on the space of closed non-empty subsets of $\mathcal{K}$.

Before we investigate the topological properties of the space of parallel classes and the space of touching pencils, we prove that the maps that assign to every point its plus- or minus-parallel class are continuous.

(5.8) Lemma: *Suppose $\mathfrak{Mi} = (\mathcal{S}, \mathcal{K})$ is a finite-dimensional connected Minkowski plane. Then the following maps are continuous:*

$$\Gamma_+ : \mathcal{S} \to \mathcal{G}_+ : s \mapsto \Gamma_+(s),$$
$$\Gamma_- : \mathcal{S} \to \mathcal{G}_- : s \mapsto \Gamma_-(s),$$
$$\Gamma : \mathcal{S} \to \mathcal{G} : s \mapsto \Gamma(s).$$

Proof: Suppose $s_i \in \mathcal{S}$ is a sequence converging to some $s \in \mathcal{S}$. We have to show that $\Gamma_+(s_i)$ converges to $\Gamma_+(s)$. So suppose $t_i \in \Gamma_+(s_i)$ is a sequence of points converging to some $t \in \mathcal{S}$. We claim $t \in \Gamma_+(s)$. Take two non-parallel points a, b such that neither of these points is parallel to any of the points s_i nor to s. Then the circles $\alpha(s_i, a, b)$ are well defined and $t = \lim_{i\to\infty} t_i = \lim_{i\to\infty} \gamma_+(s_i, \alpha(t_i, a, b)) = \gamma_+(s, \alpha(t, a, b)) \in \Gamma_+(s)$. Conversely, let $t \in \Gamma_+(s)$. Take a circle k incident with t. Then $\gamma_+(s_i, k) \in \Gamma_+(s_i)$ and $t = \gamma_+(s, k) = \lim_{i\to\infty} \gamma_+(s_i, k)$. This proves that the map $\Gamma_+ : \mathcal{S} \to \mathcal{G}_+$ is continuous. An analogous argument shows that also the map $\Gamma_- : \mathcal{S} \to \mathcal{G}_-$ is continuous.

Continuity of the map $\Gamma : \mathcal{S} \to \mathcal{G}$ follows from continuity of Γ_+ and Γ_- since $\Gamma(s) = \Gamma_+(s) \cup \Gamma_-(s)$. ☺

With the help of this lemma we can determine the topological structure of the space of parallel classes and the space $\mathcal{G}$.

(5.9) Lemma: *Suppose $\mathfrak{Mi} = (\mathcal{S}, \mathcal{K})$ is a locally compact Minkowski plane with parameter t. Then the spaces $\mathcal{G}_+$ and $\mathcal{G}_-$ are homeomorphic to circles, hence to $\mathbb{S}_t$. The space $\mathcal{G}$ is homeomorphic to $\mathcal{S}$.*

Proof: Take $k \in \mathcal{K}$. The restriction $\Gamma_+ : k \to \mathcal{G}_+ : s \mapsto \Gamma_+(s)$ is continuous. The inverse map is $\Gamma_+^{-1} : \mathcal{G}_+ \to k : g \mapsto g \cap k$. Since k is compact and since $\mathcal{G}_+$ is a Hausdorff space, the restriction $\Gamma_+ : k \to \mathcal{G}_+$ is a homeomorphism ([1]: 3.7).

The map $\Gamma : \mathcal{S} \to \mathcal{G}$ is obviously bijective, thus continuity of Γ implies that Γ is even a homeomorphism by the same argument as above. ♋

Remark: The fact that the point space can be identified in a natural way with a space of certain closed subsets of $\mathcal{S}$ will turn out to be very convenient when we introduce topologies on the lifted Lie geometry. In the case of locally compact Laguerre planes with parameter 2 there is no such identification and defining suitable topologies on the Lie geometry requires some work (cf. Chapter 3 or [41]).

If $b \in \mathcal{B}$ is a touching pencil and t a point not parallel to the support $\sigma(b)$ of b, then, by axiom M2 of a Minkowski plane, there is precisely one circle in b incident with t. We will denote this circle by $\boldsymbol{\varepsilon(t, b)}$.

(5.10) Lemma: *Suppose $\mathfrak{Mi} = (\mathcal{S}, \mathcal{K})$ is a finite-dimensional connected Minkowski plane. Then the following maps are continuous:*

$$B : \mathcal{F} \to \mathcal{B} : (s,k) \mapsto B(s,k),$$
$$\sigma : \mathcal{B} \to \mathcal{S} : b \mapsto \sigma(b),$$
$$\varepsilon : \{(t,b) \in \mathcal{S} \times \mathcal{B} \,|\, \sigma(b) \notin \Gamma(t)\} \to \mathcal{K} : (t,b) \mapsto \varepsilon(t,b).$$

Proof: First we prove that the map B is continuous. Suppose $(s_i, k_i) \in \mathcal{F}$ is a sequence converging to some flag $(s,k) \in \mathcal{F}$. We have to show that $B(s_i, k_i)$ converges to $B(s,k)$. So suppose there is a sequence of circles $l_i \in B(s_i, k_i)$ converging to some circle $l \in \mathcal{K}$. Take a point $t \in \mathcal{S}$ neither parallel to s nor to any of the points s_i. Then $l = \lim_{i\to\infty} l_i = \lim_{i\to\infty} \beta(s_i, k_i, \gamma_+(t, l_i)) = \beta(s, k, \gamma_+(t, l)) \in B(s,k)$. Conversely, let $l \in B(s,k)$. Take a point t as above. Then $\beta(s_i, k_i, t) \in B(s_i, k_i)$ and $l = \beta(s, k, t) = \lim_{i\to\infty} \beta(s_i, k_i, t)$. This proves continuity of the map $B : \mathcal{F} \to \mathcal{B}$.

Continuity of the map $\sigma : \mathcal{B} \to \mathcal{S}$ follows readily from the fact that $\mathcal{F}$ is closed in $\mathcal{S} \times \mathcal{K}$.

Next we prove that the map ε is continuous. Suppose that $b_i \in \mathcal{B}$ is a sequence converging to some $b \in \mathcal{B}$ and that $t_i \in \mathcal{S}$ is a sequence converging to a point $t \in \mathcal{S}$ such that $\sigma(b_i) \notin \Gamma(t_i)$ and $\sigma(b) \notin \Gamma(t)$. By definition of the topology on $\mathcal{B}$ there is a sequence of circles $l_i \in b_i$ converging to some circle $l \in b$. Then $\lim_{i\to\infty} \varepsilon(t_i, b_i) = \lim_{i\to\infty} \beta(\sigma(b_i), l_i, t_i) = \beta(\sigma(b), l, t) = \varepsilon(t,b)$ by continuity of β and σ. ♋

5.4 The derivation of a Minkowski plane

In this section we have a closer look at the derivation of a Minkowski plane. Our main concern will be what circles look like in the derivation and how we can decide within the derivation of a flat Minkowski plane, whether two circles are in the same connected component of the circle space or not.

Recall that if s is a point of a Minkowski plane $\mathfrak{Mi} = (\mathcal{S}, \mathcal{K})$, then the derived structure

$$\mathfrak{A}(\mathfrak{Mi}, s) = (\mathcal{S} \setminus \Gamma(s), \{k \in \mathcal{K} \mid s \in k\} \cup \mathcal{G}_+ \setminus \Gamma_+(s) \cup \mathcal{G}_- \setminus \Gamma_-(s), \in)$$

is an affine plane. The derived projective plane $\mathfrak{P}(\mathfrak{Mi}, s)$ is the projective closure of the affine plane $\mathfrak{A}(\mathfrak{Mi}, s)$. The line at infinity of $\mathfrak{P}(\mathfrak{Mi}, s)$, i.e., the additional line of $\mathfrak{P}(\mathfrak{Mi}, s)$, is denoted l_∞. The sets $\mathcal{G}_+ \setminus \Gamma_+(s)$ and $\mathcal{G}_- \setminus \Gamma_-(s)$ determine two parallel classes of lines in $\mathfrak{A}(\mathfrak{Mi}, s)$ which in turn define two points on l_∞ denoted ∞_+ and ∞_- respectively.

By construction, a circle l of $\mathfrak{Mi}$ incident with s becomes a line of the derived projective plane denoted $\bar{l}$. The line $\bar{l}$ is neither incident with ∞_+ nor with ∞_-. The line pencil through ∞_+ consists of l_∞ and all lines corresponding to plus parallel classes. Analogously for ∞_-. Let k denote a circle not incident with s. Then the corresponding set $\bar{k} = k \setminus \{\gamma_+(s,k), \gamma_+(s,k)\} \cup \{\infty_+, \infty_-\}$ forms an oval in $\mathfrak{P}(\mathfrak{Mi}, s)$. By construction, the line at infinity intersects $\bar{k}$ in precisely two points, namely ∞_+ and ∞_-. If $\mathfrak{Mi}$ is a finite-dimensional connected Minkowski plane, then $\mathfrak{P}(\mathfrak{Mi}, s)$ is a topological projective plane and for every circle k not incident with s the induced oval $\bar{k}$ is a topological oval (Chapter 2 or [49]: 3.7).

Due to the different behaviour of ovals in flat and four-dimensional planes, we have to study the derived projective planes of flat and four-dimensional Minkowski planes separately. First we investigate flat Minkowski planes.

If O is an oval in a two-dimensional projective plane, then, by Theorem 1.1, the complement of O splits into two connected components. We will call the connected component that consists of all the points which lie on some line passing O the **outside** of O and the other connected component the **inside** of O.

The circle space $\mathcal{K}$ of a flat Minkowski plane also splits into two connected components and the following Lemma enables us to decide within the derived projective plane, whether two circles belong to the same connected component or not.

(5.11) Lemma: *Let s denote a point of a flat Minkowski plane.*

a) The set $\mathcal{K}_s$ of circles incident with s splits into two connected components. For two circles k_1 and k_2 of $\mathcal{K}_s$ the following properties are equivalent:

i) The circles k_1 and k_2 lie in the same connected component of $\mathcal{K}_s$.

ii) The circles k_1 and k_2 lie in the same connected component of $\mathcal{K}$.

iii) In the derived projective plane $\mathfrak{P}(\mathfrak{Mi}, s)$ the points on l_∞ incident with $\overline{k}_1$ or $\overline{k}_2$ lie in the same connected component of $l_\infty \setminus \{\infty_+, \infty_-\}$.

b) A circle k_1 through s and a circle k_2 not incident with s lie in the same connected component of $\mathcal{K}$ if and only if in the derived projective plane $\mathfrak{P}(\mathfrak{Mi}, s)$ the point on l_∞ incident with $\overline{k}_1$ is outside $\overline{k}_2$.

c) Two circles k_1, k_2 not incident with s lie in the same connected component of $\mathcal{K}$ if and only if the connected component of $l_\infty \setminus \{\infty_+, \infty_-\}$ that lies inside $\overline{k}_1$ lies also inside $\overline{k}_2$.

Proof: a) Equivalence of the first two properties follows immediately from the identification of $\mathcal{K}$ with $PGL(2, \mathbb{R})$ introduced in Section 5.2.

Suppose that in the derived projective plane $\mathfrak{P}(\mathfrak{Mi}, s)$ the points on l_∞ incident with $\overline{k}_1$ or $\overline{k}_2$, denoted ∞_1 and ∞_2 respectively, lie in the same connected component of $l_\infty \setminus \{\infty_+, \infty_-\}$. If these points coincide, then, back in the Minkowski plane, k_1 and k_2 touch in s. Thus, since tangent pencils are connected, they lie in the same connected component of $\mathcal{K}$. If $\infty_1 \neq \infty_2$, then k_1 and k_2 intersect in a point t not parallel to s. Let $\mu : [0, 1] \to l_\infty \setminus \{\infty_+, \infty_-\}$ be a path from ∞_1 to ∞_2. The map that sends $r \in [0, 1]$ to the line in $\mathfrak{P}(\mathfrak{Mi}, s)$ connecting t with $\mu(r)$ defines a path from k_1 to k_2 in $\mathcal{K}_{s,t} \subset \mathcal{K}_t \subset \mathcal{K}$. So Assertion iii implies Assertion ii.

To see the implication still missing suppose that in the derived projective plane $\mathfrak{P}(\mathfrak{Mi}, s)$ the points on l_∞ incident with $\overline{k}_1$ or $\overline{k}_2$, denoted ∞_1 and ∞_2 respectively, lie in different connected component of $l_\infty \setminus \{\infty_+, \infty_-\}$. Then k_1 and k_2 intersect in a point t not parallel to s. The set $\mathcal{K}_{s,t}$ of circles incident with s and t splits into two connected components which lie in different connected components of $\mathcal{K}$ (Lemma 5.5). Suppose l_1 and l_2 are two circles in $\mathcal{K}_{s,t}$ that lie in different connected components. Then the points at infinity belonging to these circles lie in different connected components of $l_\infty \setminus \{\infty_+, \infty_-\}$ by the above. We may assume that the point at infinity belonging to l_1 lies in the same connected component as ∞_1. Then, again by the above, k_1 and l_1 lie in a common connected component of $\mathcal{K}$ and k_2 and l_2 lie in a common connected component of $\mathcal{K}$, hence k_1 and k_2 lie in different connected components since l_1 and l_2 do.

b) Suppose t is a point on $l_\infty \setminus \{\infty_+, \infty_-\}$ outside $\overline{k}_2$, then, in $\mathfrak{P}(\mathfrak{Mi}, s)$, there exists a line through t tangent to $\overline{k}_2$ by Theorem 1.1. This tangent corresponds to a circle g of the Minkowski plane that touches k_2. Since tangent pencils are connected, the circles g and k_2 lie in the same connected component of $\mathcal{K}$. Thus k_1 and k_2 share a common connected component if and only if k_1 and g do. However, by Part a of this lemma, the circles k_1 and g lie in a common connected component if and only if in the derivation at s their points at infinity lie in the same connected component of $l_\infty \setminus \{\infty_+, \infty_-\}$, i.e., both outside $\overline{k}_2$.

c) The proof runs much along the same lines as the proof to Part b and can be safely left to the interested reader. ☙

In a flat Minkowski plane two circles may or may not intersect. However, if the circles are contained in different connected components, then they intersect in precisely two points, as will be proved in the next corollary.

(5.12) Corollary: *Any two circles k and g of a flat Minkowski plane which lie in different connected components of $\mathcal{K}$ intersect in two points.*

Proof: Let s denote a point incident with k but not incident with g. Consider the derived plane $\mathfrak{P}(\mathfrak{Mi}, s)$. By Lemma 5.11.b the point on l_∞ incident with $\overline{k}$ lies inside $\overline{g}$. Thus, by Theorem 1.1, the intersection $\overline{k} \cap \overline{g}$ contains precisely two points. These two points correspond to the two points in $k \cap g$. ☙

Lemma 5.11 and Theorem 1.1 put us in position to determine the number of circles in a touching pencil touching a given circle.

(5.13) Lemma: *Suppose k is a circle and b is a tangent pencil of a flat Minkowski plane such that the support of b is not incident with k. If k lies in the connected component of $\mathcal{K}$ containing the circles of b, then there are precisely two circles in b touching k. If k lies in the connected component of $\mathcal{K}$ not containing the circles of b, then there is no circle in b touching k.*

Proof: The second assertion is trivial since touching pencils are connected. So suppose the circle k lies in the same connected component as the circles of b. Let $s := \sigma(b)$ denote the support of b and consider the derived projective plane $\mathfrak{P}(\mathfrak{Mi}, s)$. In the derived projective plane the touching pencil b together with the line at infinity l_∞ becomes a line pencil through some point t on l_∞. By Lemma 5.11 the point t is outside $\overline{k}$, thus, by Theorem 1.1.a, there are precisely two lines through t tangent to $\overline{k}$. Since l_∞ is not tangent to $\overline{k}$, the two tangent lines correspond to two circles in b touching k. Conversely, any circle in b touching k corresponds to a line through t tangent to $\overline{k}$. ☉

The next lemma treats the same problem in locally compact Minkowski planes with parameter 2.

(5.14) Lemma: *Suppose k is a circle and b is a tangent pencil of a locally compact Minkowski plane with parameter 2 such that the support of b is not incident with k. Then precisely two circles in b touch k.*

Proof: Let $s := \sigma(b)$ denote the support of b and consider the derived projective plane $\mathfrak{P}(\mathfrak{Mi}, s)$. The touching pencil b corresponds to a line pencil through some point $t \in l_\infty$ not incident with $\overline{k}$. By Theorem 1.1.b there are precisely two lines through t tangent to $\overline{k}$. These two tangent lines correspond to the two circles in b touching k. ☉

Remark: Lemma 5.13 and Lemma 5.14 state for finite-dimensional connected Minkowski planes what Lemma 4.1 states for flat Möbius planes. Since the results are almost identical, the further geometric treatment is almost the same. However, topologically there are many differences.

5.5 The Lie geometry of a Minkowski plane

As in the case of Möbius plane our first attempt to obtain a quadrangle from a Minkowski plane is to consider the Lie geometry of a Minkowski plane. Again, this means that points and circles are treated as objects of the same kind. Moreover, we have a relation on the set of circles and points, called collinearity. We say that two points are collinear, if they are parallel, two circles are collinear if they touch and a point is collinear with a circle if the point is incident with the circle. As before we want the Lie geometry to be a point-line geometry. Also, as in the case of Möbius planes, it is no good to define lines as maximal sets of pairwise collinear points of the Lie geometry. Instead, we take equivalence classes and extended tangent pencils as lines. More precisely, the **Lie geometry $\mathfrak{Lie}(\mathfrak{Mi})$** of a Minkowski plane $\mathfrak{Mi} = (\mathcal{S}, \mathcal{K}, \mathcal{G}_+, \mathcal{G}_-)$ is the following incidence geometry:

$$\mathfrak{Lie}(\mathfrak{Mi}) := \left(\mathcal{S} \cup \mathcal{K}, \mathcal{G}_+ \cup \mathcal{G}_- \cup \overline{\mathcal{B}}, \in\right).$$

The elements of $\mathcal{S} \cup \mathcal{K}$ are called points and the elements of $\mathcal{G}_+ \cup \mathcal{G}_- \cup \overline{\mathcal{B}}$ are called lines. One notes that for every non-incident point-line pair (p, l) of the Lie geometry of a Minkowski plane there is always precisely one point on l collinear with p, unless p is a circle of the Minkowski plane and l is an extended touching pencil such that the support of l is not incident with p. For finite-dimensional connected Minkowski planes Lemma 5.13 and Lemma 5.14 state that in this situation there is either no point on l collinear with p or there are precisely two points on l collinear with p. So the Lie geometry of a finite-dimensional connected Minkowski plane is not a quadrangle — but almost.

Again, this result is analogous to the corresponding result for flat Möbius planes. There we introduced an orientation of circles to overcome this drawback. Unfortunately, we cannot orient circles in the same fashion. The reason is that for finite-dimensional connected Minkowski planes the point space minus a circle is still connected. But we have seen that orienting circles corresponds to a twofold covering $c : \widetilde{\mathcal{K}}_{\mathfrak{Mo}} \to \mathcal{K}_{\mathfrak{Mo}}$. This gives the decisive clue for the present situation.

5.6 The lifted Lie geometry of a Minkowski plane

Let $\mathfrak{Mi} = (\mathcal{S}, \mathcal{K})$ denote a finite-dimensional connected Minkowski plane. We have seen that there exists a twofold covering $c : \widetilde{\mathcal{K}} \to \mathcal{K}$ of the circle space. We will now show that this covering induces a twofold covering of the space of (extended) touching pencils.

Let $b \in \mathcal{B}$ be a touching pencil considered as a subset of $\mathcal{K}$. Then b is homeomorphic to $\mathbb{R}$ or $\mathbb{R}^2$, in particular b is simply connected (Lemma 5.6). Thus $c^{-1}(b)$ consists of two connected components $\tilde{b}_1$ and $\tilde{b}_2$. Let $\widetilde{\mathcal{B}} := \bigcup_{b \in \mathcal{B}} \{\tilde{b}_1, \tilde{b}_2\}$ be equipped with the topology induced by the Hausdorff metric on $\widetilde{\mathcal{K}}$. Then $c : \widetilde{\mathcal{K}} \to \mathcal{K}$ induces a map $c : \widetilde{\mathcal{B}} \to \mathcal{B}$. It is easily checked that this map is a twofold covering. If $\mathfrak{Mi}$ has parameter 2, then $\widetilde{\mathcal{B}}$ is connected. If $\mathfrak{Mi}$ has parameter 1, then $\mathcal{B}$ splits into two connected components $\mathcal{B}_1$ and $\mathcal{B}_2$, where $\mathcal{B}_i = \{b \in \mathcal{B} \mid b \subset \mathcal{K}_i\}$, $i = 1, 2$. Thus $\widetilde{\mathcal{B}}$ splits also into two connected components $\widetilde{\mathcal{B}}_1$ and $\widetilde{\mathcal{B}}_2$ and there are twofold coverings $c_i : \widetilde{\mathcal{B}}_i \to \mathcal{B}_i$ induced by $c_i : \widetilde{\mathcal{K}}_i \to \mathcal{K}_i$.

We extend $c : \widetilde{\mathcal{K}} \to \mathcal{K}$ to $c : \widetilde{\mathcal{K}} \cup \mathcal{S} \to \mathcal{K} \cup \mathcal{S}$ by defining the restriction of c to $\mathcal{S}$ to be the identity. Later we will define topologies on the relevant spaces such that the map $c : \widetilde{\mathcal{K}} \cup \mathcal{S} \to \mathcal{K} \cup \mathcal{S}$ becomes a branched covering. The extended map c induces a twofold covering $c : \widetilde{\overline{\mathcal{B}}} \to \overline{\mathcal{B}}$ of the space of extended touching pencils. The careful reader notes that we have not yet defined a topology on $\widetilde{\overline{\mathcal{B}}}$. The reason is that there are several ways to put a topology on $\overline{\overline{\mathcal{B}}}$. However, one always ends up with the same topology (Lemma 5.23). So, for the moment we may define the topology on $\overline{\mathcal{B}}$ by requiring that the natural bijection $b \mapsto b \cup \sigma(b)$ from $\mathcal{B}$ to $\overline{\mathcal{B}}$ is a homeomorphism and likewise for $\widetilde{\overline{\mathcal{B}}}$. In the case of flat Minkowski planes we also obtain twofold coverings $c : \widetilde{\overline{\mathcal{B}}}_i \to \overline{\mathcal{B}}_i$, $i = 1, 2$, where $\overline{\mathcal{B}}_i = \{b \cup \sigma(b) \mid b \in \mathcal{B}_i\}$.

These twofold coverings of the circle space and the space of extended touching pencils enable us to define the **lifted Lie geometry of a finite-dimensional connected Minkowski plane** to be the following incidence geometry:

$$\widetilde{\mathfrak{Lie}}(\mathfrak{Mi}) := (\mathcal{P} := \mathcal{S} \cup \widetilde{\mathcal{K}}, \mathcal{L} := \widetilde{\overline{\mathcal{B}}} \cup \mathcal{G}_+ \cup \mathcal{G}_-, \in).$$

So the points of the lifted Lie geometry are the points of the Minkowski plane plus the elements of the twofold covering space of the circle space of the Minkowski plane. The lines of the lifted Lie geometry are the minus- and plus-parallel classes of the Minkowski plane plus the elements of the covering space of the space of extended tangent pencils of the Minkowski plane. Thus, one can view the lifted Lie geometry as a twofold branched covering of the Lie geometry; there is a twofold covering from the point space of $\widetilde{\mathfrak{Lie}}(\mathfrak{Mi})$ to the point space of $\mathfrak{Lie}(\mathfrak{Mi})$ branched at $\mathcal{S}$ and a twofold covering from the line space of $\widetilde{\mathfrak{Lie}}(\mathfrak{Mi})$ to the line space of $\mathfrak{Lie}(\mathfrak{Mi})$ branched at $\mathcal{G}_+ \cup \mathcal{G}_-$.

If $\mathfrak{Mi}$ is a flat Minkowski plane, then $\widetilde{\mathfrak{Lie}}(\mathfrak{Mi})$ splits into $\widetilde{\mathfrak{Lie}}_1(\mathfrak{Mi})$ and $\widetilde{\mathfrak{Lie}}_2(\mathfrak{Mi})$, where $\widetilde{\mathfrak{Lie}}_i(\mathfrak{Mi}) := (\mathcal{S} \cup \widetilde{\mathcal{K}}_i, \widetilde{\overline{\mathcal{B}}}_i \cup \mathcal{G}, \in)$, $i = 1, 2$.

Notation: As in the case of flat Möbius planes plane we call the elements of $\widetilde{\mathcal{K}}$ **oriented circles** or **cycles** and the elements of $\widetilde{\mathcal{B}}$ and $\widetilde{\overline{\mathcal{B}}}$ **cycle pencils** and **extended cycle pencils** respectively. The continuous maps β, γ_+, γ_-, B, $\overline{B}$ and σ of the Minkowski plane lift to continuous maps $\widetilde{\boldsymbol{\beta}}$, $\widetilde{\gamma}_+$, $\widetilde{\gamma}_-$, $\widetilde{\boldsymbol{B}}$, $\widetilde{\overline{\boldsymbol{B}}}$ and $\widetilde{\boldsymbol{\sigma}}$ respectively. Continuity of the lifted maps follows from straightforward arguments and the proof is left to the interested reader. Two points of the lifted Lie geometry are called **collinear** if they are incident with a common line.

Our aim is to show that the lifted Lie geometry of a locally compact Minkowski plane with parameter 2 and the two components of the lifted Lie geometry of a flat Minkowski plane are generalised quadrangles.

In order to prove that the lifted Lie geometry of a Minkowski plane is a quadrangle we have to show that for every non-incident point-line-pair (p, l) there is precisely one point on l collinear with p. This is almost obvious unless the point p is a cycle and the line is a cycle pencil. This situation needs some preparation and has to be treated differently for different parameters. We start the investigation with Minkowski planes with parameter 1.

(5.15) Lemma: *Suppose x_1, x_2, x_3 are three circles of a flat Minkowski plane $\mathfrak{Mi}$ not contained in a common touching pencil such that x_i touches x_j. Let $s_1 := x_2 \cap x_3$, $s_2 := x_3 \cap x_1$ and $s_3 := x_1 \cap x_2$. Then $\delta(s_1, s_2)$ and $\delta(s_1, s_3)$ lie in different connected components of $\Gamma_+(s_1) \setminus \{s_1, \gamma_+(s_1, x_1)\}$.*

Proof: The circles x_2 and x_3 touch in s_1, thus in the derived projective plane $\mathfrak{P}(\mathfrak{Mi}, s_1)$ these circles correspond to lines $\overline{x}_2$ and $\overline{x}_3$ respectively incident with a point t on l_∞ different from ∞_+ and ∞_-. So $\overline{x}_2$ and $\overline{x}_3$ are the two lines through t tangent to $\overline{x}_1$. They divide the line pencil $\mathcal{L}_t$ through t into two connected components, one of which contains all lines through t intersecting $\overline{x}_1$ in two points and the other contains all lines through t passing $\overline{x}_1$. Since l_∞ is a line through t intersecting $\overline{x}_1$ in two points, namely ∞_+ and ∞_-, the points s_2 and s_3 lie in different connected components of $\overline{x}_1 \setminus \{\infty_+, \infty_-\}$.

The line pencil $\mathcal{L}_{\infty_-}$ through ∞_- consists of l_∞ and all lines that correspond to minus-parallel classes of $\mathfrak{Mi}$. In particular, the lines $p_i := s_i \vee \infty_-$ correspond to $\Gamma_-(s_i)$, where $i = 2, 3$. The unique line g through ∞_- tangent to $\overline{x}_1$ corresponds to $\Gamma_-(\gamma_+(s_1, x_1))$. The space $\mathcal{L}_{\infty_-} \setminus \{g\}$ is homeomorphic to $\mathbb{R}$ and the map $\iota : \overline{x}_1 \setminus \{\infty_-\} \to \mathcal{L}_{\infty_-} \setminus \{g\} : u \mapsto u \vee \infty_-$ is a homeomorphism. Since s_2 and s_3 lie in different connected components of $\overline{x}_1 \setminus \{\infty_+, \infty_-\}$, the lines $p_2 = \iota(s_2)$ and $p_3 = \iota(s_3)$ lie in different connected components of $\mathcal{L}_{\infty_-} \setminus \{g, l_\infty = \iota(\infty_+)\}$.

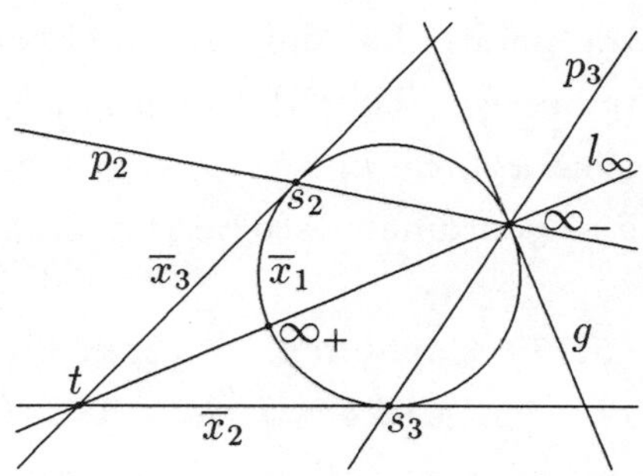

Fig. 5.1

Back in the Minkowski plane the above result states that the punctured parallel classes $\Gamma_-(s_2) \setminus \{\delta(s_1, s_2)\}$ and $\Gamma_-(s_3) \setminus \{\delta(s_1, s_3)\}$, considered as sets of points, lie in different connected components of $\mathcal{S} \setminus \big(\Gamma(s_1) \cup \Gamma_-(\gamma_+(s_1, x_1))\big)$. This is equivalent to the statement that $\Gamma_-(s_2)$ and $\Gamma_-(s_3)$ lie in different connected components of $\mathcal{G}_- \setminus \{\Gamma_-(s_1), \Gamma_-(\gamma_+(s_1, x_1))\}$. This implies the desired result since the map that sends every minus-parallel class to its intersection with $\Gamma_+(s_1)$ is a homeomorphism by Lemma 5.9. ♋

(5.16) Lemma: *Suppose b_1, b_2, b_3 are three touching pencils of a flat Minkowski plane such that any two of the touching pencils have precisely one circle in common.*

Then $b_1 \cup b_2 \cup b_3$ generates the fundamental group of one of the components $\mathcal{K}_i$ of $\mathcal{K}$, i.e., the injection $\iota : b_1 \cup b_2 \cup b_3 \hookrightarrow \mathcal{K}$ induces an isomorphism $\iota_* : \pi_1(b_1 \cup b_2 \cup b_3) \to \pi_1(\mathcal{K}_i)$.

Proof: Let $x_1 := b_2 \cap b_3$, $x_2 := b_3 \cap b_1$ and $x_3 := b_1 \cap b_2$. The touching pencils b_i, $i = 1, 2, 3$, are homeomorphic to $\mathbb{R}$ and the circles x_2, x_3, x_1 determine closed intervals $I_i \subset b_i$. The set $b_1 \cup b_2 \cup b_3$ is homotopic to $\Sigma := I_1 \cup I_2 \cup I_3$ and the latter is homeomorphic to $\mathbb{S}_1$. In particular $b_1 \cup b_2 \cup b_3$ is connected and hence contained in one of the connected components of $\mathcal{K}$. Let $s_i := \sigma(b_i)$ and $t := \gamma_+(s_1, x_1)$. We claim that the points s_i are mutually non-parallel. By construction $\{s_i, s_j\} \in x_k$ where $\{i, j, k\} = \{1, 2, 3\}$. Thus s_i is not parallel to s_j, unless $s_i = s_j$. In this case $b_i = B(s_i, x_k) = B(s_j, x_k) = b_j$, contradicting the assumptions. Hence s_i is not parallel to s_j. For I_1 the set $\gamma_+(s_1, I_1)$ consists just of the point s_1, for I_i, where $i \in \{2, 3\}$, the set $\gamma_+(s_1, I_i)$ is the closure of the connected component of $\Gamma_+(s_1) \setminus \{s_1, t\}$ not containing $\delta(s_1, s_i,)$. By Lemma 5.15 the points $\delta(s_1, s_2,)$ and $\delta(s_1, s_3,)$ lie in different connected components of $\Gamma_+(s_1) \setminus \{s_1, t\}$, thus $\gamma_+(s_1, \Sigma) = \Gamma_+(s_1)$. Furthermore, the point t has only one preimage — in fact all points different from s_1 have one preimage. Thus the map $\kappa : b_1 \cup b_2 \cup b_3 \to \Gamma_+(s_1) : x \mapsto \gamma_+(s_1, x)$ induces an isomorphism $\kappa_* : \pi_1(b_1 \cup b_2 \cup b_3) \to \pi_1(\Gamma_+(s_1))$. Coordinatising $\mathcal{K}$ using $\Gamma(s_1)$ as first coordinate (see Section 5.2) proves that the assertion is true. ♋

(5.17) Corollary: *Suppose $\widetilde{\mathfrak{Lie}}_i(\mathfrak{Mi})$ is one component of the lifted Lie geometry of a flat Minkowski plane $\mathfrak{Mi}$. Let $p \in \widetilde{\mathcal{K}}_i$ and $l \in \widetilde{\mathcal{B}}_i$ such that the support of l is not incident with p. Then precisely one cycle in l touches p and precisely one cycle pencil containing p intersects l.*

Proof: Consider $x_1 := c(p)$ and $b_1 := c(l)$. Then by Lemma 5.5 precisely two circles x_2 and x_3 of the touching pencil b_1 touch x_1. Let b_2 denote the touching pencil spanned by x_1 and x_3 and let b_3 denote the touching pencil spanned by x_1 and x_2. By Lemma 5.16 the space $b_1 \cup b_2 \cup b_3$ is a fundamental circle of $\mathcal{K}_i$, thus the restriction $c : c^{-1}(b_1 \cup b_2 \cup b_3) \to b_1 \cup b_2 \cup b_3$ is again a twofold covering. In particular, the space $c^{-1}(b_1 \cup b_2 \cup b_3)$ is connected. Let q_j denote the element of $c^{-1}(x_j)$ contained in l, where $j = 2, 3$. Then precisely one of the cycles q_2 and q_3 touches p. For otherwise, if both touch p, then the preimage $c^{-1}(b_1 \cup b_2 \cup b_3)$ would split into two connected components. Likewise, if none of the points q_2 and q_3 touches p, then both touch the element of $c^{-1}(x_1)$ different from p which again yields a contradiction to the fact that $c^{-1}(b_1 \cup b_2 \cup b_3)$ is connected.

The cycle pencil spanned by p and the unique cycle on l touching p is the unique cycle pencil containing p that intersects l. ☺

We now prove an analogous result for the parameter 2 case.

(5.18) Lemma: *Suppose $l \in \widetilde{\mathcal{B}}$ is a cycle pencil and p a cycle of a locally compact Minkowski plane with parameter 2 such that p is not incident with the support of l. Then precisely one cycle in l touches p and precisely one cycle pencil containing p intersects l.*

Proof: Let $s := \sigma(l)$ and $k := c(p)$. First we will see that the map that assigns to each touching pencil b containing k the unique touching pencil with support s that intersects b is a twofold covering. We then will see that there exists an induced map from the space of cycle pencils containing p to the space of cycle pencils with support s and that this map is a homeomorphism. This then proves the lemma. In this proof it is necessary to make a clear distinction between a circle or cycle x and the set $\mathcal{S}_x$ of points incident with x.

The space $\mathcal{B}_k := \{b \in \mathcal{B} \,|\, k \in b\}$ is naturally homeomorphic to the space $\mathcal{S}_k := \{x \in \mathcal{S} \,|\, x \in k\}$ of points incident with k via the map $\sigma_k : \mathcal{B}_k \to \mathcal{S}_k : b \mapsto \sigma(b)$ whose inverse map is $\sigma_k^{-1} : \mathcal{S}_k \to \mathcal{B}_k : x \mapsto B(x,k)$. For a cycle p let $\widetilde{\mathcal{B}}_p := \{b \in \widetilde{\mathcal{B}} \,|\, k \in b\}$. Then $\widetilde{\sigma}_p : \widetilde{\mathcal{B}}_p \to \mathcal{S}_p : b \mapsto \widetilde{\sigma}(b)$ is also a homeomorphism with inverse map $\widetilde{\sigma}_p^{-1} : \mathcal{S}_p \to \widetilde{\mathcal{B}}_p : x \mapsto \widetilde{B}(x,p)$.

Let $t \in \mathcal{S}$ be any point of $\mathfrak{Mi}$ not parallel to s. Then the space $\mathcal{K}_{s,t}$ is homeomorphic to $\mathcal{B}_s := \{b \in \mathcal{B} \,|\, \sigma(b) = s\}$ via the map $B_{s,t} : \mathcal{K}_{s,t} \to \mathcal{B}_s : z \mapsto B(s,z)$ whose inverse map is $B_{s,t}^{-1} : \mathcal{B}_s \to \mathcal{K}_{s,t} : b \mapsto \varepsilon(t,b)$. Again, the homeomorphism $B_{s,t}$ induces a homeomorphism $\widetilde{B}_{s,t} : \widetilde{\mathcal{K}}_{s,t} \to \widetilde{\mathcal{B}}_s : z \mapsto \widetilde{B}(s,k)$ where $\widetilde{\mathcal{B}}_s := \{b \in \widetilde{\mathcal{B}} \,|\, \sigma(b) = s\}$.

Let $\mathcal{S}'_k := \mathcal{S}_k \setminus \Gamma(s)$ and consider the continuous map

$$\varphi : \mathcal{S}'_k \to \mathcal{B}_s : x \mapsto B(s, \beta(x,k,s)).$$

We want to prove that this map is a twofold covering. We argue in the derived affine plane $\mathfrak{A}(\mathfrak{Mi}, s)$. There the map φ corresponds to the map that sends every point x of $k \setminus \{\gamma_+(s,k), \gamma_-(s,k)\}$ to the set of lines parallel to the tangent at x. So, in the derived projective plane $\mathfrak{P}(\mathfrak{Mi}, s)$, the map φ induces the map $\overline{\varphi}$ that sends every point $x \in \overline{k} \setminus \{\infty_+, \infty_-\}$ to the intersection of the tangent at x with l_∞. Now, the set of lines tangent to $\overline{k}$ is again a topological oval ([4]: 3.8), which says that the map that sends every point y of $\mathfrak{P}(\mathfrak{Mi}, s)$ to the set of lines through y tangent to $\overline{k}$ is continuous. From this we can deduce that the map $\overline{\varphi}$ is locally injective. Moreover, since precisely two lines through a point not incident with $\overline{k}$ are tangent to $\overline{k}$ ([4]: 3.9), the map $\overline{\varphi}$ is a twofold covering. This in turn implies that also the map φ is a twofold covering as claimed.

What about the map $\widetilde{\varphi} : \mathcal{S}'_p \to \widetilde{\mathcal{K}}_{s,t} : x \mapsto \widetilde{B}(s, \widetilde{\beta}(x,p,s))$ induced by φ, where $\mathcal{S}'_p := \{q \in \mathcal{S} \setminus \Gamma(s) \,|\, q \in p\}$? We claim that $\widetilde{\varphi}$ is a homeomorphism. The restriction of c to $\mathcal{S}$ is the identity, thus the restriction $c : \mathcal{S}'_p \to \mathcal{S}'_k$ is also the identity, i.e., $\mathcal{S}_p = \mathcal{S}_k$. Furthermore, by Lemma 5.6, the restriction $c : \widetilde{\mathcal{K}}_{s,t} \to \mathcal{K}_{s,t}$ is a twofold covering with a connected covering space. It induces a twofold covering $c : \widetilde{\mathcal{B}}_s \to \mathcal{B}_s$ with connected covering space $\widetilde{\mathcal{B}}_s$. Finally, the map φ is a twofold covering by the above, so the map $\widetilde{\varphi} : \mathcal{S}_p \to \widetilde{\mathcal{B}}_s$ is a onefold covering, i.e., a homeomorphism. The following diagram, where $\mathcal{B}'_p := \{b \in \mathcal{B} \,|\, \sigma(b) \not\parallel s\}$ and $\widetilde{\mathcal{B}}'_p := \{b \in \widetilde{\mathcal{B}} \,|\, \widetilde{\sigma}(p) \not\parallel s\}$, sketches the situation.

$$\begin{array}{ccccccc}
\widetilde{\mathcal{B}}'_p & \xrightarrow{\widetilde{\sigma}_p} & \mathcal{S}'_p & \xrightarrow{\widetilde{\varphi}} & \widetilde{\mathcal{B}}_s & \xleftarrow{\widetilde{B}_{s,t}} & \widetilde{\mathcal{K}}_{s,t} \\
\downarrow{\scriptstyle c} & & \downarrow{\scriptstyle c} & & \downarrow{\scriptstyle c} & & \downarrow{\scriptstyle c} \\
\mathcal{B}'_k & \xrightarrow{\sigma_k} & \mathcal{S}'_k & \xrightarrow{\varphi} & \mathcal{B}_s & \xleftarrow{B_{s,t}} & \mathcal{K}_{s,t}
\end{array}$$

The map $\widetilde{\varphi} \circ \widetilde{\sigma}_p : \widetilde{\mathcal{B}}'_p \to \widetilde{\mathcal{B}}_s$ assigns to every cycle pencil $b \in \widetilde{\mathcal{B}}_p$ whose support is not parallel to s the unique cycle pencil in $\widetilde{\mathcal{B}}_s$ intersecting with b. It is a homeomorphism since $\widetilde{\varphi}$ and $\widetilde{\sigma}_p$ are. Thus $(\widetilde{\varphi} \circ \widetilde{\sigma}_p)^{-1}(l) = \widetilde{B}(\widetilde{\varphi}^{-1}(l), p)$ is the unique cycle pencil that contains p and intersects l. Moreover, the cycle $\widetilde{\beta}(\widetilde{\varphi}^{-1}(l), p, s)$ is the unique element of l touching p.

We are now in a position to prove that we can construct generalised quadrangles from finite-dimensional connected Minkowski plane via the lifted Lie geometries.

(5.19) Theorem: *The lifted Lie geometry $\widetilde{\mathfrak{Lie}}(\mathfrak{Mi})$ of a locally compact Minkowski plane with parameter 2 and the two components $\widetilde{\mathfrak{Lie}}_1(\mathfrak{Mi})$ and $\widetilde{\mathfrak{Lie}}_2(\mathfrak{Mi})$ of the lifted Lie geometry of a locally compact Minkowski plane with parameter 1 are generalised quadrangles.*

Proof: Let $\mathcal{P}$ denote the point space and let $\mathcal{L}$ denote the line space of either the lifted Lie geometry of a locally compact Minkowski plane with parameter 2 or one of the two components of the lifted Lie geometry of a locally compact Minkowski plane with parameter 1. We have to show that for every pair $(p,l) \in \mathcal{P} \times \mathcal{L}$ with $p \notin l$ there exists precisely one point $\pi(p,l)$ in l collinear with p and one line $\lambda(p,l)$ through p

intersecting l. The point p can be a point or a cycle and the line l can be a plus- or minus-parallel class or a cycle pencil. We have to treat these cases separately.

- $p \in \mathcal{S}$, $l \in \mathcal{G}$: If l is a plus-parallel class, i.e., if $l = \Gamma_+(t)$ for some t, then $\pi(p,l) = \delta(t,p) = \Gamma_-(p) \cap l$ and $\lambda(p,l) = \Gamma_-(p)$. If l is a minus-parallel class, i.e., if $l = \Gamma_-(t)$ for some t, then $\pi(p,l) = \delta(p,t) = \Gamma_+(p) \cap l$ and $\lambda(p,l) = \Gamma_+(p)$.
- $p \in \mathcal{S}$, $l \in \widetilde{\widetilde{\mathcal{B}}}$: Let $s := \widetilde{\sigma}(l)$. If $s \in \Gamma_+(p)$ or $s \in \Gamma_-(p)$, then $\pi(p,l) = s$ and $\lambda(p,l) = \Gamma_+(p)$ or $\lambda(p,l) = \Gamma_-(p)$ respectively. If $s \not\parallel p$ then $\pi(p,l) = \widetilde{\beta}(s,l,p)$ and $\lambda(p,l) = \widetilde{\widetilde{B}}(p,\pi(p,l))$.
- $p \in \widetilde{\mathcal{K}}$, $l \in \mathcal{G}$: In this case $\pi(p,l)$ is the point in l incident with p, i.e., if $l = \Gamma_+(t)$ or $l = \Gamma_-(t)$ for some t, then $\pi(p,l) = \widetilde{\gamma}_+(t,p)$ or $\pi(p,l) = \widetilde{\gamma}_-(t,p)$ respectively and $\lambda(p,l) = \widetilde{\widetilde{B}}(\pi(p,l),p)$.
- $p \in \widetilde{\mathcal{K}}$, $l \in \widetilde{\widetilde{\mathcal{B}}}$: If $\widetilde{\sigma}(l) \in p$, then $\pi(p,l) = \widetilde{\sigma}(l)$ and $\lambda(p,l) = \widetilde{\widetilde{B}}(\widetilde{\sigma}(l),p)$. If the support of l is not incident with p, then there is a unique cycle in l touching p and a unique extended cycle pencil containing p and intersecting l by Corollary 5.17 if the parameter of the Minkowski plane is 1 and Lemma 5.18 if the parameter of the Minkowski plane is 2. So, $\pi(p,l)$ is this unique point and $\lambda(p,l)$ is this unique line. ۞

Remark: The above proof shows some resemblance with the proof of the analogous theorems for topological Laguerre planes (Theorem 3.4 or [37] and [41]) and for flat Möbius planes (Theorem 4.4). In both cases the topological requirements were only needed for the situation where p is an (oriented) circle and l an extended (oriented) touching pencil whose support is not incident with p. Moreover, the above proof with the exception of the last situation holds also for the Lie geometry of a Minkowski plane, justifying our earlier remark that the Lie geometry of a Minkowski plane is almost a generalised quadrangle.

Notation: From now on the lifted Lie geometry of a locally compact Minkowski plane $\mathfrak{Mi}$ with parameter 2 will be denoted $\mathfrak{GQ}(\mathfrak{Mi})$ and the two components of the lifted Lie geometry of a locally compact Minkowski plane $\mathfrak{Mi}$ with parameter 1 will be denoted $\mathfrak{GQ}_1(\mathfrak{Mi})$ and $\mathfrak{GQ}_2(\mathfrak{Mi})$.

5.7 The topology on the lifted Lie geometry

In this section we want to show that the generalised quadrangles constructed from finite-dimensional connected Minkowski planes can be made topological quadrangles in a rather natural way. We first equip the point space of the Lie geometry with a topology such that it becomes compact and metrisable. Then we define a topology on the line space of the Lie geometry and show that with this topology the line space becomes a compact space as well. Finally, we prove that with these topologies the lifted Lie geometry of a finite-dimensional connected Minkowski plane consists of topological quadrangles.

We have seen in Lemma 5.7 and Proposition 5.1 that parallel classes as well as circles are closed sets. So, it is natural to define the topology on $\mathcal{G} \cup \mathcal{K}$ using the Hausdorff metric. This extends the topologies already defined on $\mathcal{K}$ and $\mathcal{G}$.

(5.20) Lemma: *In a finite-dimensional connected Minkowski plane $\mathfrak{Mi}$ the space $\mathcal{G} \cup \mathcal{K}$ is compact with regard to the topology induced by the Hausdorff metric.*

Proof: The space $\mathcal{G}$ is homeomorphic to $\mathcal{S}$ by Lemma 5.9, hence compact. Thus all we have to show is that for every sequence $k_n \in \mathcal{K}$, $n \in \mathbb{N}$, of circles there exists a subsequence which converges to a circle or to some element of $\mathcal{G}$. Let (s_1, s_2, s_3) and (t_1, t_2, t_3) denote two triples of pairwise non-parallel points. Taking a subsequence of the sequence k_n, $n \in \mathbb{N}$, if necessary, we may assume that for every $i = 1, 2, 3$ the sequence $\gamma_+(s_i, k_n)$, $n \in \mathbb{N}$, converges to some point u_i and the sequence $\gamma_-(t_i, k_n)$, $n \in \mathbb{N}$, converges to some point v_i.

The sequence k_n, $n \in \mathbb{N}$, converges to some circle $k \in \mathcal{K}$ if and only if the points u_1, u_2 and u_3 are mutually non-parallel. In this case the points v_1, v_2 and v_3 are also mutually non-parallel and $\lim_{n\to\infty} k_n = k = \alpha(u_1, u_2, u_3) = \alpha(v_1, v_2, v_3)$.

If the sequence k_n, $n \in \mathbb{N}$, has no accumulation point in $\mathcal{K}$, then at least two of the points u_1, u_2 and u_3 as well as two of the points v_1, v_2 and v_3 are parallel. We may assume $u_1 \parallel_- u_2$ and $v_1 \parallel_+ v_2$. Moreover, we may assume $v_2 \nparallel u_2$. However, the points v_1 and u_1 might even coincide. Let $x := \delta(v_2, u_1)$. We claim that $\lim_{n\to\infty} k_n = \Gamma(x)$.

We know already $\{u_1, u_2, v_1, v_2\} \subset \Gamma(x)$. The set $\{u_1, u_2, u_3, v_1, v_2, v_3\}$ contains no triple of pairwise non-parallel points for otherwise the sequence k_n, $n \in \mathbb{N}$, would converge to the circle spanned by such a triple in contradiction to our assumption. Combinatorial arguments show that if the set $\{u_1, u_2, u_3, v_1, v_2, v_3\}$ contains no triple of pairwise non-parallel points, then it is contained in $\Gamma(x)$. Now suppose $z_n \in k_n$, $n \in \mathbb{N}$, is a sequence converging to some point $z \in \mathcal{S}$. If $z \notin \Gamma(x)$, we may assume $u_2 \nparallel z \nparallel v_2 \nparallel u_2$. Thus $\lim_{n\to\infty} k_n = \lim_{n\to\infty} \alpha(\gamma_+(s_2, k_n), \gamma_-(t_2, k_n), z_n) = \alpha(u_2, v_2, z) \in \mathcal{K}$ contradicting our assumption. Hence $z \in \Gamma(x)$ and therefore $\lim_{n\to\infty} k_n \subset \Gamma(x)$.

Next suppose $y \in \Gamma_+(x) \setminus \{x\}$. Compactness of $\Gamma_-(y)$ implies that the sequence $\gamma_-(y, k_n) \in k_n$, $n \in \mathbb{N}$, converges to some point $y' \in \Gamma_-(y)$. Thus $y' \in \Gamma_-(y) \cap \lim_{n\to\infty} k_n \subset \Gamma_-(y) \cap \Gamma(x) = \{y\}$, i.e., $y' = y$. Hence $y \in \lim_{n\to\infty} k_n$ and therefore $\Gamma_+(x) \setminus \{x\} \subset \lim_{n\to\infty} k_n$. An analogous argument shows that also $\Gamma_-(x) \setminus \{x\} \subset \lim_{n\to\infty} k_n$, hence $\Gamma(x) \setminus \{x\} \subset \lim_{n\to\infty} k_n$.

If the Minkowski has parameter 2, then every circle l through x meets each of the circles k_n, $n \in \mathbb{N}$. If the Minkowski has parameter 1, then, by Corollary 5.12, there exists a circle l through x meeting infinitely many of the circles k_n, $n \in \mathbb{N}$. So, taking a subsequence if necessary, we may assume that in any case there is a circle l through x such that for every $n \in \mathbb{N}$ the intersection $l \cap k_n$ contains at least one point x_n. Compactness of the circle l and the fact that $\lim_{n\to\infty} k_n \subset \Gamma(x)$ imply $\lim_{n\to\infty} x_n = x$. Thus $\Gamma(x) \subset \lim_{n\to\infty} k_n$, hence, with the above, $\Gamma(x) = \lim_{n\to\infty} k_n$ as claimed. ☙

Extending the identification of $\mathcal{S}$ with $\mathcal{G}$ established in Lemma 5.9 to a identification of $\mathcal{S} \cup \mathcal{K}$ with $\mathcal{G} \cup \mathcal{K}$ defines a topology on $\mathcal{S} \cup \mathcal{K}$. By the above Lemma the space $\mathcal{S} \cup \mathcal{K}$ is compact. By definition, the map $c : \widetilde{\mathcal{K}} \to \mathcal{K}$ is a twofold covering. Requiring the map $c : \mathcal{S} \cup \widetilde{\mathcal{K}} \to \mathcal{S} \cup \mathcal{K}$ to be continuous defines a topology on $\mathcal{S} \cup \widetilde{\mathcal{K}}$. With this topology the space $\mathcal{S} \cup \widetilde{\mathcal{K}}$ is compact.

To verify that the space $\mathcal{S} \cup \widetilde{\mathcal{K}}$ is metrisable we proceed as in the case of flat Möbius planes (Section 4.3). If a flat Minkowski plane is given, choose four pairwise non-parallel points s_i, $i \in \{1, 2, 3, 4\}$, not contained in a common circle. Then, as in Section 4.3 one sees that the subspace of $\mathcal{S} \cup \widetilde{\mathcal{K}}$ consisting of points not contained in U_i and cycles missing U_i is metrisable. This then induces that $\mathcal{S} \cup \widetilde{\mathcal{K}}$ is metrisable. If a locally compact Minkowski plane with parameter 2 is given, we have to proceed in a slightly different way. The reason is that the space of cycles missing a given point is still connected. We therefore identify the circle space with the group $PSL(2, \mathbb{C})$ and the space of cycles with the group $SL(2, \mathbb{C})$ such that the covering $c : \widetilde{\mathcal{K}} \to \mathcal{K}$ is equivalent to the covering $P : SL(2, \mathbb{C}) \to PSL(2, \mathbb{C})$. Let a_i, $i \in \{1, 2, 3, 4\}$ denote the entries of a two by two matrix. For $i \in \{1, 2, 3, 4\}$ let K_i denote those matrices in $SL(2, \mathbb{C})$ with $|\mathrm{Im}(a_i)| < 1/4$ and let Λ_i denote those matrices in $SL(2, \mathbb{C})$ with $|\mathrm{Re}(a_i)| < 1/4$. Then $\bigcap_{i=1}^4 (\mathrm{K}_i \cap \Lambda_i) = \emptyset$. Also, $SL(2, \mathbb{C}) \setminus \mathrm{K}_i$ as well as $SL(2, \mathbb{C}) \setminus \Lambda_i$ consists of two connected components each of which is homeomorphic to $\mathbb{C}^2 \times \mathbb{R} \times \{t \in \mathbb{R} \mid t \geq 0\}$. Moreover, $P(SL(2, \mathbb{C}) \setminus \mathrm{K}_i) = PSL(2, \mathbb{C}) \setminus P(\mathrm{K}_i)$ is connected and via P homeomorphic to each component of $SL(2, \mathbb{C}) \setminus \mathrm{K}_i$. The same holds for $SL(2, \mathbb{C}) \setminus \Lambda_i$. The space $\mathcal{S} \cup PSL(2, \mathbb{C}) \setminus P(\mathrm{K}_i)$, considered as a subset of $\mathcal{S} \cup \mathcal{K}$, is metrisable, thus $\mathcal{S} \cup X$, where X denotes a connected component of $SL(2,\mathbb{C})\setminus \mathrm{K}_i$ considered as a subset of $\mathcal{S}\cup\widetilde{\mathcal{K}}$, is also metrisable. By [9]: 4.4.19 this implies that $\mathcal{S}\cup SL(2,\mathbb{C})\setminus \mathrm{K}_i$ is metrisabel. Likewise, one verifies that $\mathcal{S}\cup SL(2,\mathbb{C})\setminus \Lambda_i$ is metrisable. A further application of [9]: 4.4.19 implies that $\bigcup_{i=1}^4((\mathcal{S}\cup SL(2,\mathbb{C})\setminus \mathrm{K}_i)\cup(\mathcal{S}\cup SL(2,\mathbb{C})\setminus \Lambda_i)) = \mathcal{S}\cup SL(2,\mathbb{C})\setminus\bigcap_{i=1}^4(\mathrm{K}_i\cap\Lambda_i) = \mathcal{S}\cup SL(2,\mathbb{C}) = \mathcal{S}\cup\widetilde{\mathcal{K}}$ is metrisable.

The topology on $\mathcal{S}\cup\widetilde{\mathcal{K}}$ can be described in terms of sequences as follows. On the subspaces $\mathcal{S}$ and $\widetilde{\mathcal{K}}$ the topology coincides with the original topology and a sequence of cycles $\widetilde{k}_n$, $n\in\mathbb{N}$, converges to a point $s\in\mathcal{S}$ if and only if the sequence $c(\widetilde{k}_n)$, $n\in\mathbb{N}$, converges to s, i.e., if and only if the sequence $c(\widetilde{k}_n)$, $n\in\mathbb{N}$, converges to $\Gamma(s)$ with regard to the topology induced by the Hausdorff metric. One checks also readily that if $\mathfrak{Mi}$ is a flat Minkowski plane, then the spaces $\mathcal{S}\cup\widetilde{\mathcal{K}}_1$ and $\mathcal{S}\cup\widetilde{\mathcal{K}}_2$ are compact as well.

We summarise this in the following corollary:

(5.21) Corollary: *For every finite-dimensional connected Minkowski plane $\mathfrak{Mi} = (\mathcal{S},\mathcal{K})$ the point space $\mathcal{S}\cup\mathcal{K}$ of the Lie geometry as well as the point space $\mathcal{S}\cup\widetilde{\mathcal{K}}$ of the lifted Lie geometry is compact and metrisable. Moreover, if $\mathfrak{Mi}$ is a flat Minkowski plane, the spaces $\mathcal{S}\cup\mathcal{K}_i$ as well as the spaces $\mathcal{S}\cup\widetilde{\mathcal{K}}_i$, $i=1,2$, are compact and metrisable.*

We now turn to the line space $\overline{\mathcal{B}}\cup\mathcal{G}_+\cup\mathcal{G}_-$ of the Lie geometry of a finite-dimensional Minkowski plane. First we define a topology on the space $\overline{\mathcal{B}}$ of extended tangent pencils. Since we want to work mainly with the topology induced by the Hausdorff metric on the space $\mathcal{K}\cup\mathcal{G}$, we have to view every extended tangent pencil as a closed subset of $\mathcal{K}\cup\mathcal{G}$.

(5.22) Lemma: *For every tangent pencil $b\in\mathcal{B}$ of a finite-dimensional connected Minkowski plane $\mathfrak{Mi}$ the space $b\cup\Gamma(\sigma(b))$ is closed in $\mathcal{K}\cup\mathcal{G}$.*

Proof: By Lemma 5.7.d every tangent pencil b is closed in $\mathcal{K}$ and by Lemma 5.20 the space $\mathcal{K}\times\mathcal{G}$ is compact. So, all we have to show is that if a sequence $k_i\in b$, $i\in\mathbb{N}$, converges to $\Gamma(t)$ for some $t\in\mathcal{S}$, then $t=\sigma(b)$. Suppose $t\neq\sigma(b)$. Choose a point $x\in\Gamma(t)$ not parallel to $\sigma(b)$ and fix a circle l in b. By the definition of the topology on $\mathcal{K}$ there is a sequence of points $x_i\in k_i$, $i\in\mathbb{N}$, converging to x. Without loss of generality we may assume that none of the points x_i is parallel to $\sigma(b)$. Then $k_i=\beta(s,l,x_i)$, thus $\lim_{i\to\infty}k_i=\lim_{i\to\infty}\beta(s,l,x_i)=\beta(s,l,x)\in\mathcal{K}$ contradicting our assumptions.

Identifying every extended touching pencil $b \cup \sigma(b) \in \overline{\mathcal{B}}$ with the set $b \cup \Gamma(\sigma(b)) \in \mathcal{K} \cup \mathcal{G}$ we can view every extended touching pencil as a closed subset of $\mathcal{K} \cup \mathcal{G}$. Therefore, the Hausdorff metric on $\mathcal{K} \cup \mathcal{G}$ induces a topology on $\overline{\mathcal{B}}$. We will denote this topology by T_1 for the moment.

We can also view every extended touching pencil $b \cup \{\sigma(b)\}$ as a closed subset of the disjoint union $\mathcal{K} \dot{\cup} \mathcal{S}$. Therefore, the Hausdorff metrics on $\mathcal{K}$ and $\mathcal{S}$ define another topology T_2 on $\overline{\mathcal{B}}$. One checks readily that the map $s \mapsto \{s\}$ is a homeomorphism from $\mathcal{S}$ to the space of subsets of $\mathcal{S}$ containing precisely one element. Thus in the definition of the topology T_2 we can also consider the set $b \cup \sigma(b)$ rather than $b \cup \{\sigma(b)\}$.

Fortunately the topologies T_1 and T_2 are equivalent and the space of extended touching pencils equipped with one of these topologies is naturally homeomorphic to the space of touching pencils as will reveal the next lemma.

(5.23) Lemma: *Suppose $\overline{\mathcal{B}}$ is the space of extended touching pencils of a finite-dimensional connected Minkowski plane and T_1 and T_2 are topologies on $\overline{\mathcal{B}}$ as defined above. Then:*

a) The map $\varphi : (\overline{\mathcal{B}}, T_2) \to (\overline{\mathcal{B}}, T_1) : b \cup \sigma(b) \mapsto b \cup \Gamma(\sigma(b))$ is a homeomorphism.
b) The map $\psi : \mathcal{B} \to (\overline{\mathcal{B}}, T_2) : b \mapsto b \cup \sigma(b)$ is a homeomorphism.

Proof: The assertion of Part b follows readily from the continuity of the map σ and the definition of the topology T_2. Almost the same arguments as in the proof of Lemma 5.22 imply continuity of the map φ. Continuity of the inverse map is obvious. ♋

Since the topologies T_1 and T_2 are equivalent, we will no longer specify explicitly which one we use. For technical reasons we will mainly work with the first one.

It follows from Lemma 5.22 and the definition of the topology on $\widetilde{\mathcal{K}} \cup \mathcal{G}$ that extended cycle pencils are closed in $\widetilde{\mathcal{K}} \cup \mathcal{G}$. Thus we can equip $\overline{\widetilde{\mathcal{B}}}$ with the topology induced by the Hausdorff metric. As above, one verifies that $\overline{\widetilde{\mathcal{B}}}$ is naturally homeomorphic to $\widetilde{\mathcal{B}}$. Thus the map $c : \overline{\widetilde{\mathcal{B}}} \to \overline{\mathcal{B}}$ is a twofold covering.

Our next aim is to define a topology on the line space $\overline{\mathcal{B}} \cup \mathcal{G}_+ \cup \mathcal{G}_-$ of the Lie geometry of a finite-dimensional connected Minkowski plane. Let

$$T := \{\{\Gamma(s) \mid s \in l\} \mid l \in \mathcal{G}_+ \cup \mathcal{G}_-\}$$

be equipped with the topology induced by the Hausdorff metric on $\mathcal{G}$. By Lemma 5.8 the map $\Gamma : \mathcal{S} \to \mathcal{G} : s \mapsto \Gamma(s)$ is a homeomorphism, thus the space T is naturally homeomorphic to $\mathcal{G}_+ \cup \mathcal{G}_-$. The elements of $\overline{\mathcal{B}} \cup T$ are closed subsets of $\mathcal{K} \cup \mathcal{G}$. Therefore, the Hausdorff metric defines a topology on $\overline{\mathcal{B}} \cup T$. This topology extends the given topologies on $\overline{\mathcal{B}}$ and T. Identifying T with $\mathcal{G}_+ \cup \mathcal{G}_-$ we obtain a topology on $\overline{\mathcal{B}} \cup \mathcal{G}_+ \cup \mathcal{G}_-$ extending the topologies on $\overline{\mathcal{B}}$, $\mathcal{G}_+$ and $\mathcal{G}_-$.

(5.24) Lemma: *The line space $\overline{\mathcal{B}} \cup \mathcal{G}_+ \cup \mathcal{G}_-$ of the Lie geometry of a finite-dimensional connected Minkowski plane $\mathfrak{Mi}$ is compact.*

Proof: Since the space $\mathcal{G}_+ \cup \mathcal{G}_- \cong T$ is compact, we only have to prove that if $g_i \in \overline{\mathcal{B}}$, $i \in \mathbb{N}$, is a sequence of extended touching pencils having no accumulation point in $\overline{\mathcal{B}}$, then there is a subsequence converging to some $g \in \mathcal{G}_+ \cup \mathcal{G}_-$.

Let $s_i := \sigma(g_i)$. By compactness of $\mathcal{S}$ we may assume that the sequence s_i, $i \in \mathbb{N}$, converges to some $s \in \mathcal{S}$. Suppose there are circles $k_i \in g_i$ converging to some circle $k \in \mathcal{K}$. Then, by Lemma 5.10, $\lim_{i\to\infty} g_i = \lim_{i\to\infty} \overline{\mathcal{B}}(s_i, k_i) = \overline{\mathcal{B}}(s, k) \in \overline{\mathcal{B}}$ contradicting our assumptions. The space $\mathcal{K} \cup \mathcal{G}$ is compact, thus the accumulation points of every sequence $k_i \in g_i$, $i \in \mathbb{N}$, lie in $\mathcal{G}$.

Let $t \in \mathcal{S} \setminus \Gamma(s)$. We may assume that also $t \notin \Gamma(s_i)$ for every $i \in \mathbb{N}$. Let $k_i := \varepsilon(t, g_i)$ denote the unique circle in g_i incident with t. By compactness of $\mathcal{K} \cup \mathcal{G}$ we may assume that there is some $v \in \mathcal{S}$ such that $\lim_{i\to\infty} k_i = \Gamma(v)$. Now $\{t, s_i\} \in k_i$ for every $i \in \mathbb{N}$, thus $\{t, s\} \in \Gamma(v)$. Therefore $v \in \Gamma(s) \cap \Gamma(t) = \{\delta(s,t), \delta(t,s)\}$.

We claim that $\lim_{i\to\infty} k_i = \Gamma(s) \cap \Gamma(v)$, i.e., $\lim_{i\to\infty} k_i = \Gamma_+(s) = \Gamma_+(v) \in \mathcal{G}_+$ if $v = \delta(s,t)$ and $\lim_{i\to\infty} k_i = \Gamma_-(s) = \Gamma_-(v) \in \mathcal{G}_-$ if $v = \delta(t,s)$. We will treat only the case $\lim_{i\to\infty} k_i = \Gamma_+(s)$, the other case is analogous.

Suppose $l_i \in g_i$, $i \in \mathbb{N}$, is some convergent sequence of circles. By the above, we know that there is some $w \in \mathcal{S}$ such that $\lim_{i\to\infty} l_i = \Gamma(w)$. Let $z_i := \gamma_-(v, l_i) \in \Gamma_-(v)$. Since $\Gamma_-(v)$ is compact we may assume that there is some $z \in \Gamma_-(v)$ such that $\lim_{i\to\infty} z_i = z$. If $z = v$, then $s, v \in \Gamma(w)$ thus $w \in \Gamma(v) \cap \Gamma(s) = \Gamma_+(s)$ as claimed. If $z \neq v$, then, as above, either $w = \delta(s,z) = v \in \Gamma_+(s)$ as claimed or $w \in \delta(z,s) \in \Gamma_-(s)$. We have to exclude the second case.

To exclude the case $w \in \delta(z,s)$ we identify $\mathcal{S}$ with $\Gamma_+(s) \times \Gamma_-(s)$. Let $N_v \subset M_v$ denote two neighbourhoods of v in $\Gamma_+(s)$ and let $N_w \subset M_w$ denote two neighbourhoods of w in $\Gamma_-(s)$, such that every neighbourhood is homeomorphic to the closed unit disk of the appropriate dimension, such that N_v and N_w are contained in the interior of M_v and M_w respectively and such that neither M_v nor M_w contains s.

Let $M_z := M_v \times M_w$ and let R_z denote the boundary of M_z. Since $\lim_{i\to\infty} k_i = \Gamma(v)$ and $\lim_{i\to\infty} l_i = \Gamma(w)$ there is some $I \in \mathbb{N}$ such that $k_I \cap (\Gamma_+(s) \times M_w) \subset$

$\mathrm{int}(N_v) \times M_w$ and $l_I \cap (M_v \times \Gamma_-(s)) \subset M_v \times \mathrm{int}(N_w)$ where $\mathrm{int}(X)$ denotes the interior of X. In particular $k_I \neq l_I$.

If $\mathfrak{Mi}$ is a flat Minkowski plane, then R_z is an ordinary quadrangle with vertices (a_i, b_j), $i, j \in \{1, 2\}$, where $\{a_1, a_2\}$ is the boundary of M_v and $\{b_1, b_2\}$ is the boundary of M_w. The intersection $k_I \cap R_z$ consists of two points $c_i \in N_v \times \{b_i\} \subset M_v \times \{b_i\}$, $i = 1, 2$ and the intersection $l_I \cap R_z$ consists of two points $d_i \in \{a_i\} \times N_w \subset \{a_i\} \times M_w$, $i = 1, 2$. The line $k_I \cap M_z$ connects c_1 to c_2 and splits M_z into two connected components each of which contains one of the points d_1 and d_2. However, the line $l_I \cap M_z$ connects d_1 to d_2, thus $l_I \cap k_I \cap M_z$ is not empty. Let $y \in l_I \cap k_I \cap M_z$. Then k_I is the unique line in g_I containing y and l_I is the unique line in g_I containing y, i.e., $k_I = l_I$, contradicting the choice of $I \in \mathbb{N}$.

If the Minkowski plane $\mathfrak{Mi}$ has parameter 2, the arguments run along similar lines, replacing connectedness arguments by arguments involving fundamental groups. Let R_v denote the boundary of M_v and let R_w denote the boundary of M_w. Then $R_v \cong R_w \cong \mathbb{S}_1$. The space $M_z \setminus \Gamma_-(v)$ is homeomorphic to $\mathbb{R}^2 \times (\mathbb{R}^2 \setminus \{(0,0)\})$, in particular its fundamental group is the infinity cyclic group. The intersection $\Gamma_+(w) \cap R_z$ generates the fundamental group of $M_z \setminus \Gamma_-(v)$. The restriction $\gamma_{+,k_I} : \Gamma_-(v) \cap M_z \to k_I \cap M_z : r \mapsto \gamma_+(r, k_I)$ is a homeomorphism and can be extended to a homeomorphism from $N_v \times M_w$ onto itself which leaves the sets $\Gamma_+(r) \cap M_z$ fixed for every $r \in M_w$ and which is the identity map on $\partial N_v \times M_w$, where ∂N_v denotes the boundary of N_v. Furthermore, the restriction $\gamma_{-,l_I} : \Gamma_+(w) \cap R_z \to l_I \cap R_z : r \mapsto \gamma_-(r, l_I)$ is a homeomorphism and can be extended to a homeomorphism from $M_z \setminus k_I$ onto itself. Therefore, the intersection $l_I \cap R_z$ generates the non-trivial fundamental group of $M_z \setminus k_I$. Thus $l_I \cap k_I \cap M_z$ is not empty, for otherwise $l_I \cap R_z$ is contractable in $l_I \cap M_z \subset M_z \setminus k_I$. As in the flat case, the fact that $l_I \cap k_I \cap M_z$ is not empty yields a contradiction. So, for both possible parameters we have excluded the case $w \in \Gamma_-(s)$.

We still assume $v \in \Gamma_+(s)$. Up to now we know $\lim_{i\to\infty} g_i \subset \Gamma_+(s)$. Let $z \in \Gamma_+(s) \setminus \{s\}$ and take $t \in \Gamma_-(z) \setminus \{z\}$. Moreover, let $l_i := \varepsilon(t, g_i)$ denote the unique circle in g_i containing t. Then, as above, we see that the sequence l_i, $i \in \mathbb{N}$, converges to $z = \delta(s,t)$ or to $\delta(t,s)$. However, the same arguments as above show that $z \in \Gamma(s) \cap \Gamma(v) = \Gamma_+(s)$. Hence $\Gamma_+(s) \setminus \{s\}$ is contained in $\lim_{i\to\infty} g_i$. Since $s = \lim_{i\to\infty} s_i$, we know $\Gamma_+(s) \subset \lim_{i\to\infty} g_i$, thus, with the above, $\Gamma_+(s) = \lim_{i\to\infty} g_i$ as claimed. ☺

We now turn to the line space $\overline{\overline{\mathcal{B}}} \cup \mathcal{G}_+ \cup \mathcal{G}_-$ of the lifted Lie geometry. We have already a topology on $\overline{\overline{\mathcal{B}}}$ and, as in the case of $\overline{\mathcal{B}}$, the topology on $\overline{\overline{\mathcal{B}}} \cup T$ induced by the Hausdorff metric on $\widetilde{\mathcal{K}} \cup \mathcal{G}$ defines a topology on $\widetilde{\widetilde{\mathcal{B}}} \cup \mathcal{G}_+ \cup \mathcal{G}_-$, where $T := \{\{\Gamma(s) \mid s \in l\} \mid l \in \mathcal{G}_+ \cup \mathcal{G}_-\}$ as above. Compactness of $\overline{\mathcal{B}} \cup \mathcal{G}_+ \cup \mathcal{G}_-$ implies compactness of $\widetilde{\widetilde{\mathcal{B}}} \cup \mathcal{G}_+ \cup \mathcal{G}_-$. Thus we get the following corollary:

(5.25) Corollary: *The line space $\widetilde{\widetilde{\mathcal{B}}} \cup \mathcal{G}_+ \cup \mathcal{G}_-$ of the lifted Lie geometry of a finite-dimensional connected Minkowski plane $\mathfrak{Mi}$ is compact.* ֍

We are now ready to prove the main result of this section, namely that the lifted Lie geometry of a finite-dimensional connected Minkowski plane consists of topological quadrangles.

(5.26) Theorem: *Suppose $\mathfrak{Mi}$ is a locally compact Minkowski plane with parameter t. If $t = 2$, the lifted Lie geometry $\mathfrak{GQ}(\mathfrak{Mi})$ is a compact quadrangle with parameter 2. If $t = 1$, the two components $\mathfrak{GQ}_1(\mathfrak{Mi})$ and $\mathfrak{GQ}_2(\mathfrak{Mi})$ of the lifted Lie geometry $\mathfrak{GQ}(\mathfrak{Mi})$ are compact quadrangles with parameter 1.*

Proof: By Theorem 5.19 the geometries in question are generalised quadrangles. Let $\mathcal{P}$ denote the point space and let $\mathcal{L}$ denote the line space of one of these quadrangles. By Corollary 5.21 and Corollary 5.25 the point space and the line space of the lifted Lie geometry of a finite-dimensional connected Minkowski plane are compact. It follows that also the point spaces and the line spaces of the quadrangles constructed from the lifted Lie geometry are compact. Furthermore, the space $\mathcal{F} \in \mathcal{P} \times \mathcal{L}$ of incident point-line pairs is closed in $\mathcal{P} \times \mathcal{L}$, since we have defined the topology on $\mathcal{L}$ with the Hausdorff metric on $\mathcal{P}$. Therefore, if $(p_i, l_i) \in \mathcal{F}$, $i \in \mathbb{N}$, is a sequence converging to some $(p, l) \in \mathcal{P} \times \mathcal{L}$, then $p \in l$ by the definition of the topology on $\mathcal{L}$, thus $(p, l) \in \mathcal{F}$. Let $\mathcal{D} := \mathcal{P} \times \mathcal{L} \setminus \mathcal{F}$ denote the domain of definition of the map $\lambda \times \pi$. Then, by the above, the graph $\{(q, g, l, z) \in \mathcal{D} \times \mathcal{L} \times \mathcal{P} \mid \{(q, l), (z, l), (z, q)\} \subset \mathcal{F}\}$ is closed in $\mathcal{D} \times \mathcal{L} \times \mathcal{P}$. Therefore $\lambda \times \pi$ is continuous ([8]: XI.2.7), thus the maps λ and π are also continuous.

Suppose the Minkowski plane has parameter t. Then, by Corollary 5.2, every element of $\mathcal{G}$ is homeomorphic to $\mathbb{S}_t$. Let $k \in \widetilde{\mathcal{K}}$. Then $\varphi : c(k) \to \mathcal{L}_k : r \mapsto \widetilde{B}(r, k)$ is a homeomorphism from $c(k) \cong \mathbb{S}_t$ to the space $\mathcal{L}$ of lines incident with k. Since in a thick topological quadrangle any two point rows and any two line pencils are homeomorphic ([11]: 2.4), we can deduce that every point row as well as every line pencil is homeomorphic to $\mathbb{S}_t$. ֍

Remark: As in the case of Möbius planes one can also define maps π and λ on the Lie geometry $\mathfrak{Lie}(\mathfrak{Mi})$ itself. For every non-incident point-line pair (p, l) we say that $\pi(p,l)$ consists of all points on l incident with p and that $\lambda(p,l)$ consists of all lines through p meeting l. We have seen that $\pi(p,l)$ and $\lambda(p,l)$ are either both empty or contain both one element or contain both two elements. Restricting the domain of definition we can exclude the cases were the image is the empty set. Then the Hausdorff metric defines a topology on the space of sets consisting of one or two points and the space of sets consisting of one or two lines. One sees readily that with these topologies the maps π and λ are continuous. So we can say that the Lie geometry of a finite-dimensional connected Minkowski plane is a topological incidence geometry.

5.8 Characterising quadrangles obtained from Minkowski planes

In this section we characterise the quadrangles and the pairs of quadrangles obtained as the lifted Lie geometry of a locally compact Minkowski plane with parameters 2 or 1 respectively. Moreover, we will show that given such a quadrangle or such a pair of quadrangles one can reconstruct the Minkowski plane. More precisely, we will see that the quadrangles obtained from locally compact Minkowski planes with parameters t are compact antiregular quadrangles with parameter t and certain additional properties. We also find a way to construct a flat Minkowski plane from just one compact quadrangle with parameter 1.

We begin our considerations with the flat case.

(5.27) Lemma: *The quadrangles $\mathfrak{G}\mathfrak{Q}_1(\mathfrak{Mi})$ and $\mathfrak{G}\mathfrak{Q}_2(\mathfrak{Mi})$, where $\mathfrak{Mi}$ is a flat Minkowski plane, are antiregular.*

Proof: By Theorem 5.26 we can view the quadrangles as compact quadrangles with parameter 1. Let k denote a circle of $\mathcal{K}_i$, $i \in \{1,2\}$, and let s_1, s_2, s_3 denote three points incident with k. Then the three points are mutually non-parallel, thus the intersection $\Gamma(s_1) \cap \Gamma(s_2) \cap \Gamma(s_3)$ is empty. Moreover, there is no circle in the other connected component $\mathcal{K}_{3-i}$ incident with s_1, s_2 and s_3. Thus there is no point in the lifted Lie geometry $\mathfrak{G}\mathfrak{Q}_{3-i}(\mathfrak{Mi})$ incident with all of the points s_1, s_2, s_3. By Lemma 2.17 this implies that the quadrangle $\mathfrak{G}\mathfrak{Q}_{3-i}(\mathfrak{Mi})$ is antiregular. ♋

The twofold covering $c : \widetilde{\mathcal{K}} \to \mathcal{K}$ comes with a covering transformation τ on $\widetilde{\mathcal{K}}$, i.e., a continuous fixed point free involution τ such that $c(p) = c(\tau(p))$ for every cycle $p \in \widetilde{\mathcal{K}}$. We extend this map by the identity on $\mathcal{S}$ to the whole point space of the lifted Lie geometry. This extension is continuous since a sequence $p_i \in \widetilde{\mathcal{K}}$, $i \in \mathbb{N}$, of cycles converges to a point $s \in \mathcal{S}$ if and only if the corresponding sequence $c(p_i)$, $i \in \mathbb{N}$, converges to s. The map τ induces an involution on the line space of the lifted Lie geometry. The lines fixed by this involution are precisely the lines in $\mathcal{G}_+ \cup \mathcal{G}_-$. So the map τ defines a continuous involutory collineation on the lifted Lie geometry $\mathfrak{GQ}(\mathfrak{Mi})$ of the flat Minkowski plane $\mathfrak{Mi}$. We denote this collineation by $\tau_{\mathfrak{Mi}}$. The induced collineations on the components $\mathfrak{GQ}_i(\mathfrak{Mi})$ are denoted $\tau_{i,\mathfrak{Mi}}$, $i = 1, 2$. The set of points and the set of lines fixed by $\tau_{i,\mathfrak{Mi}}$ form a non-thick generalised subquadrangle $\mathcal{V}_i$ of $\mathfrak{GQ}_i$ where point rows are homeomorphic to $\mathbb{S}_1$ and where line pencils contain precisely two elements, i.e., where line pencils are homeomorphic to $\mathbb{S}_0$. Thus, with notation introduced in Section 2.1, we obtain a compact quadrangle with parameters $(1,0)$.

If s_1, s_2, s_3 are three mutually non-parallel points of the Minkowski plane, then the connecting circle $\alpha(s_1, s_2, s_3)$ is contained in one of the connected components of $\mathcal{K}$, hence the two cycles incident with s_1, s_2 and s_3 lie in the same connected component of $\widetilde{\mathcal{K}}$. Thus in one of the components $\mathfrak{Lie}_i(\mathfrak{Mi})$ of the lifted Lie geometry there are precisely two points collinear with all of the points s_1, s_2, s_3 and in the other component there are no points collinear with all of the points s_1, s_2 and s_3. These considerations prove the next result.

(5.28) Theorem: *Suppose $\mathfrak{GQ}_i = \mathfrak{GQ}_i(\mathfrak{Mi})$, $i = 1, 2$, are the two components of the lifted Lie geometry of a flat Minkowski plane $\mathfrak{Mi}$. Then for $i = 1$ and 2 the quadrangles $\mathfrak{GQ}_i$ are antiregular and compact with parameter 1 and there are continuous involutory collineations τ_i on $\mathfrak{GQ}_i$ such that the space of fixed points and the space of fixed lines form a compact subquadrangle $\mathcal{V}_i$ with parameters $(1,0)$. Moreover, there is a continuous isomorphism $\varphi : \mathcal{V}_1 \to \mathcal{V}_2$ such that $|s_1^\perp \cap s_2^\perp \cap s_3^\perp| = 2 - |\varphi(s_1)^\perp \cap \varphi(s_2)^\perp \cap \varphi(s_3)^\perp|$ for any triad $\{s_1, s_2, s_3\}$ in $\mathcal{V}_1$.* ☺

As for the other types of circle geometries the above construction can be reversed. The reversed construction works also in the non-topological case. The generalisation of compact quadrangles with parameters $(t, 0)$ are **point thin quadrangles**. These are quadrangles where every point is incident with precisely two lines.

(5.29) Theorem: *Suppose $\mathfrak{GQ}_1 = (\mathcal{P}_1, \mathcal{L}_1)$ and $\mathfrak{GQ}_2 = (\mathcal{P}_2, \mathcal{L}_2)$ are two thick non-centric antiregular quadrangles, suppose τ_i are involutory collineations on $\mathfrak{GQ}_i$ such that $\mathcal{V}_i := (\Sigma_i, \Lambda_i, \in)$, where Σ_i denotes the fixed point set of τ and Λ_i the fixed line set of τ_i, are point thin quadrangles and suppose there is an isomorphism*

$\varphi : \mathcal{V}_1 \to \mathcal{V}_2$ such that $|s_1^\perp \cap s_2^\perp \cap s_3^\perp| = 2 - |\varphi(s_1)^\perp \cap \varphi(s_2)^\perp \cap \varphi(s_3)^\perp|$ for any triad $\{s_1, s_2, s_3\}$ in $\mathcal{V}_1$. Then the derived structure

$$\mathfrak{Mi}(\mathfrak{GQ}_1, \mathfrak{GQ}_2, \tau_1, \tau_2, \varphi) := \\ (\Sigma_1, \{\Sigma_1 \cap q^\perp \,|\, q \in \mathcal{P}_1 \setminus \Sigma_1\} \cup \{\varphi^{-1}(\Sigma_2 \cap q^\perp) \,|\, q \in \mathcal{P}_2 \setminus \Sigma_2\}, \Lambda_1, \in),$$

is a Minkowski plane.

Proof: Let $K_1 := \{\Sigma_1 \cap q^\perp \,|\, q \in \mathcal{P}_1 \setminus \Sigma_1\}$ and $K_2 := \{\varphi^{-1}(\Sigma_2 \cap q^\perp) \,|\, q \in \mathcal{P}_2 \setminus \Sigma_2\}$. Moreover, let $\kappa_1 : \mathcal{P}_1 \setminus \Sigma_1 \to K_1 : q \mapsto q^\perp \cap \Sigma_1$ and let $\kappa_2 : \mathcal{P}_1 \setminus \Sigma_2 \to K_2 : q \mapsto \varphi^{-1}(q^\perp \cap \Sigma_2)$. The set Λ_1 splits into two parts Λ_{1+} and Λ_{1-} such that every line in Λ_{1+} meets every line in Λ_{1-} but any two lines in Λ_{1+} and any two lines in Λ_{1-} do not intersect. These two sets may be constructed as follows. Take a point $s \in \Sigma_1$ and let l_+ and l_- denote the two lines of Λ_1 incident with s. Then Λ_{1+} is the set of lines in $\Lambda_1 \setminus \{l_-\}$ intersecting l_- and Λ_{1-} is the set of lines in $\Lambda_1 \setminus \{l_+\}$ intersecting l_+.

In anticipation of the claimed result we will call the elements of $K := K_1 \cup K_2$ circles, the elements of Λ_{1+} plus-parallel classes and the elements of Λ_{1-} minus-parallel classes. We see already that the point space Σ_1 is isomorphic to $\Lambda_{1+} \times \Lambda_{1-}$.

Next we show that any three mutually non-collinear points in Σ_1 define a unique circle in K. So suppose $s_1, s_2, s_3 \in \Sigma_1$ form a triad. If $s_1^\perp \cap s_2^\perp \cap s_3^\perp$ contains precisely two points p_1 and p_2, then $\{p_1, p_2\} \cap \Sigma_1 = \emptyset$, since every point in Σ_1 is contained only in two lines of Λ_1. Thus $\tau_1(p_1) = p_2$, since τ_1 is a collineation and acts without fixed points on $\mathcal{P}_1 \setminus \Sigma_1$. Hence $\{s_1, s_2, s_3\} \in \kappa_1(p_1)$ and $\kappa_1(p_1) = \Sigma_1 \cap p_1^\perp = \tau(\Sigma_1 \cap p_1^\perp) = \tau(\Sigma_1) \cap \tau(p_1^\perp) = \Sigma_1 \cap \tau(p_1)^\perp = \kappa_1(p_2) \in K_1$ is the unique element $\alpha(s_1, s_2, s_3)$ of K_1 incident with s_1, s_2, s_3. Since $|s_1^\perp \cap s_2^\perp \cap s_3^\perp| = 2$, the centre $\varphi(s_1)^\perp \cap \varphi(s_2)^\perp \cap \varphi(s_3)^\perp$ is empty, thus no circle in K_2 is incident with s_1, s_2 and s_3. If $s_1 s_2 s_3$ does not contain precisely two points, then $s_1^\perp \cap s_2^\perp \cap s_3^\perp = \emptyset$ by antiregularity. But then the centre $\varphi(s_1)^\perp \cap \varphi(s_2)^\perp \cap \varphi(s_3)^\perp$ contains precisely two points p_1 and p_2. Arguments analogous to the above show that $\alpha(s_1, s_2, s_3) := \kappa_2(p_1) = \kappa_2(p_2)$ is the unique element of K incident with s_1, s_2, s_3.

The arguments show also that for $k \in K_1$ we have $k = p_1^\perp \cap p_2^\perp$ where $\{p_1, p_2\} = \kappa_1^{-1}(k)$ and that for $k \in K_2$ we have $k = \varphi^{-1}(p_1^\perp \cap p_2^\perp)$ where $\{p_1, p_2\} = \kappa_2^{-1}(k)$.

Next we want to define and discuss the map β which is supposed to be the touching map as in the definition of a Minkowski plane. Suppose $s, t \in \Sigma_1$, $k \in K_1$, such that $s \in k$ and such that t is not collinear with s. Take $p \in \kappa_1^{-1}(k)$ and let $g \in \mathcal{L}$ denote the line joining s to p. Since t is not collinear with s, the point $\pi(t, g)$ is not contained in Σ_1. Moreover, $\tau_1(g)$ is the line joining s to $\tau_1(p)$ and

$\beta(s,k,t) := \kappa_1(\pi(t,g)) = \kappa_1(\pi(t,\tau_1(g)) \in K_1$ is the unique element of K incident with t and touching k in s. If $k \in K_2$ take $p \in \kappa_2^{-1}(k)$ and let g denote the line in $\mathcal{L}_2$ joining $\varphi(s)$ to p. Then $\beta(s,k,t) := \kappa_2(\pi(\varphi(t),g)) = \kappa_2(\pi(\varphi(t),\tau_2(g)) \in K_2$.

Our next step is to investigate parallel projections. Suppose $s \in \Sigma_1$ and $k \in K_i$. Let $g_+ \in \Lambda_{1+}$ denote the unique plus-parallel class incident with s, let $g_- \in \Lambda_{1-}$ denote the unique minus-parallel class incident with s and let $p \in \kappa_1^{-1}(k)$. Then $\gamma_+(s,k) := \pi(p,g_+)$ is the unique point on k plus-parallel to s and $\gamma_-(s,k) := \pi(p,g_-)$ is the unique point on k minus-parallel to s. If $k \in K_2$, we set $\gamma_+(s,k) := \varphi^{-1}(\pi(p,\varphi(g_+))$ and $\gamma_+(s,k) := \varphi^{-1}(\pi(p,\varphi(g_-))$ where $p \in \kappa_2^{-1}(k)$.

We now discuss the map δ mentioned in the axioms for a Minkowski plane. Suppose $s,t \in \Sigma_1$ are two non-collinear points. Let l denote the line in Λ_{1+} incident with s. Then $\delta(s,t) := \pi(t,l)$ is the unique point plus-parallel to s and minus-parallel to v.

Thickness of the quadrangles $\mathfrak{GQ}_i$ implies that the derived geometry is not degenerate. Thus the derived structure is indeed a Minkowski plane. ∎

We now return to the flat case again.

(5.30) Theorem: *Suppose $\mathfrak{GQ}_i = \mathfrak{GQ}_i(\mathfrak{Mi})$, $i = 1,2$, are the two components of the lifted Lie geometry of a flat Minkowski plane $\mathfrak{Mi}$. Then for $i = 1$ and 2 the quadrangles $\mathfrak{GQ}_i$ are antiregular and compact with parameter 1 and there are continuous involutory collineations τ_i on $\mathfrak{GQ}_i$ such that the space of fixed points and the space of fixed lines form a compact subquadrangle $\mathcal{V}_i$ with parameters $(1,0)$. Moreover, there is a continuous isomorphism $\varphi : \mathcal{V}_1 \to \mathcal{V}_2$ such that $|s_1^\perp \cap s_2^\perp \cap s_3^\perp| = 2 - |\varphi(s_1)^\perp \cap \varphi(s_2)^\perp \cap \varphi(s_3)^\perp|$ for any triad $\{s_1,s_2,s_3\}$ in $\mathcal{V}_1$.*

Conversely, if $\mathfrak{GQ}_1 = (\mathcal{P}_1,\mathcal{L}_1)$ and $\mathfrak{GQ}_2 = (\mathcal{P}_2,\mathcal{L}_2)$ are two antiregular compact quadrangles with parameter 1 and maps τ_1, τ_2 and φ as above, then the derived structure

$$\mathfrak{Mi}(\mathfrak{GQ}_1,\mathfrak{GQ}_2,\tau_1,\tau_2,\varphi) =$$
$$\left(\Sigma_1, \{\Sigma_1 \cap q^\perp \,|\, q \in \mathcal{P}_1 \setminus \Sigma_1\} \cup \{\varphi^{-1}(\Sigma_2 \cap q^\perp) \,|\, q \in \mathcal{P}_2 \setminus \Sigma_2\}, \Lambda_+ \cup \Lambda_-, \in\right),$$

where Σ_i denotes the space of points fixed by τ_i, $i = 1,2$, and where Λ_+ and Λ_- denote the two connected components of the space of lines fixed by τ_1, is a flat Minkowski plane.

Proof: The first part follows readily from Theorem 5.28. By Theorem 5.29 the derived structure is a Minkowski plane. Thus we only have to check that this Minkowski plane is a flat Minkowski plane.

For this purpose we employ the following result by A. Schenkel ([35]: 4.4 and 3.12): A Minkowski plane whose point set is homeomorphic to $\mathbb{S}_1 \times \mathbb{S}_1$ and where circles as well as plus- and minus-parallel classes are homeomorphic to $\mathbb{S}_1$ becomes a flat Minkowski plane if we equip the circle space with the topology induced by the Hausdorff metric on the space of closed subsets of the point space.

The spaces Λ_+ and Λ_- correspond to the spaces Λ_{1+} and Λ_{1-} of Theorem 5.29. Thus Λ_+ and Λ_- are naturally homeomorphic to lines of $\mathfrak{G}\mathfrak{Q}_1$, hence to $\mathbb{S}_1$. The point space Σ_1 is homeomorphic to $\Lambda_+ \times \Lambda_-$ thus to $\mathbb{S}_1 \times \mathbb{S}_1$.

If s_1, s_2 and s_3 are three mutually non-collinear points in Σ_1, then $\alpha(s_1, s_2, s_3) = p^\perp \cap \tau_1(p)^\perp$ if $s_1^\perp \cap s_2^\perp \cap s_3^\perp = \{p, \tau_1(p)\} \neq \emptyset$ or $\alpha(s_1, s_2, s_3) = \varphi^{-1}(p^\perp \cap \tau_2(p)^\perp)$ if $\varphi(s_1)^\perp \cap \varphi(s_2)^\perp \cap \varphi(s_3)^\perp = \{p, \tau_2(p)\} \neq \emptyset$. Thus circles are homeomorphic to line pencils, hence to $\mathbb{S}_1$.

Therefore all topological requirements needed to employ the above mentioned result by A. Schenkel hold. Thus the derived structure is even a flat Minkowski plane, if the topology on K is defined via the Hausdorff metric. ✪

Remarks: One could also prove that the derived Minkowski plane is topological using the fact that certain maps in the generalised quadrangles $\mathfrak{G}\mathfrak{Q}_1$ and $\mathfrak{G}\mathfrak{Q}_2$ are continuous. We will take this way in the parameter 2 case. However, to prove that in the flat case the intersection of two circles is continuous we cannot use continuous maps of the generalised quadrangles if the circles belong to two different connected components. In this case one has to work within the Minkowski plane using similar arguments as A. Schenkel in her proof of the result employed above. In the parameter 2 case this situation does not occur since there is only one generalised quadrangle and the circle space is connected. Moreover, I do not know of an analogy of A. Schenkel's result for the parameter 2 case.

A natural question is which pairs of antiregular quadrangles lead to a Minkowski plane. In Theorem 5.29 we showed that any pair of thick, non-centric antiregular quadrangles $\mathfrak{G}\mathfrak{Q}_i$ with involutions τ_i subject to certain conditions and with a collineation φ between the fixed geometries yields a Minkowski plane. The point of interest is, whether such a collineation always exist. We make no attempt to answer this question in the general case. One certainly has to impose some other conditions on the quadrangles. But in the flat case we can answer the question in the affirmative. More precisely we have the following result:

(5.31) Theorem: *Suppose $\mathfrak{G}\mathfrak{Q}_1 = (\mathcal{P}_1, \mathcal{L}_1)$ and $\mathfrak{G}\mathfrak{Q}_2 = (\mathcal{P}_2, \mathcal{L}_2)$ are two antiregular compact quadrangles with parameter 1 and suppose τ_i are involutory collineations* *on* $\mathfrak{G}\Omega_i$ *such that* $\mathcal{V}_i := (\Sigma_i, \Lambda_i, \in)$*, where* Σ_i *denotes the fixed point set of* τ_i *and* Λ_i *the fixed line set of* τ_i*, are compact quadrangles with parameters* $(1,0)$*. Then there exists a continuous isomorphism* $\varphi : \mathcal{V}_1 \to \mathcal{V}_2$ *such that* $|s_1^\perp \cap s_2^\perp \cap s_3^\perp| = 2 - |\varphi(s_1)^\perp \cap \varphi(s_2)^\perp \cap \varphi(s_3)^\perp|$ *for any triad* $\{s_1, s_2, s_3\}$ *in* $\mathcal{V}_1$*.*

Proof: Let Λ_{i+} and Λ_{i-} denote the two connected components of Λ_i. As in the proof of the preceding theorem one sees that each of these connected components is homeomorphic to $\mathbb{S}_1$. For $i \in \{1,2\}$ we fix an orientation of Λ_{i+} and an orientation of Λ_{i-}.

For every point $p \in \mathcal{P}_1 \setminus \Sigma_1$ we obtain a map $\psi_p : \Lambda_{i+} \to \Lambda_{i+}$ which maps $l \in \Lambda_{i+}$ to the unique line in Λ_{i-} incident with $\pi(p,l)$. This map is continuous as the following argument shows: Let $g \in \Lambda_{i+}$. Then, for $l \in \Lambda_{i+} \setminus \{g\}$ we have $\psi_p(l) = \lambda(\pi(p,l), g)$. Thus ψ_p is continuous on $\Lambda_{i+} \setminus \{g\}$. Since g was an arbitrary element of Λ_{i+}, the map ψ_p is continuous on the whole of Λ_{i+}. The inverse map is of the same fashion. It assigns to every line l of Λ_{i-} the unique line in Λ_{i+} incident with $\pi(p,l)$. As above one verifies that this map is also continuous, thus ψ_p is a homeomorphism.

The point of interest is now, whether or nor ψ_p is an orientation preserving or an orientation reversing map. We claim that either ψ_p is orientation preserving for every point $p \in \mathcal{P}_1 \setminus \Sigma_1$ or ψ_p is orientation reversing for every point $p \in \mathcal{P}_1 \setminus \Sigma_1$.

To prove this claim suppose p is a point such that ψ_p preserves orientation. Then there is a neighbourhood U_p of p in $\mathcal{P}_1 \setminus \Sigma_1$ such that ψ_q is orientation preserving for every $q \in U_p$. To obtain such a neighbourhood choose three lines l_1, l_2 and l_3 in Λ_{1+}. Let $s_i := \pi(p, l_i)$ for $i \in \{1,2,3\}$. Then the points s_i are pairwise non-collinear. Thus, there exist connected neighbourhoods U_i of s_i in l_i such that for $i \neq j$ no point in U_i is collinear with a point in U_j. Let $U_p := \{q \in \mathcal{P}_1 \setminus \Sigma_1 \mid \pi(q, l_i) \in U_i;\ i \in \{1,2,3\}\}$. Then, by construction, ψ_q preserves orientation since ψ_p does so. It remains to show that U_p is a neighbourhood of p.

Consider the square projection $\mu_{s_1,s_2,l_3} : s_1^\perp \cap s_2^\perp \to l_3$. By construction U_3 is contained in a connected component of $l_3 \setminus \{\pi(s_1, l_3), \pi(s_2, l_3)\}$. Thus, since $s_3 = \mu_{s_1,s_2,l_3}(p)$ is in the image of the square projection, so is U_3. Therefore $\mu^{-1}_{s_1,s_2,l_3}(U_3)$ consists of two connected components and the restriction of the square projection to each component is a homeomorphism onto U_3. Applying analogous arguments first to square projections μ_{s_1,t_3,l_2}, where $t_3 \in U_3$, and then to square projections μ_{t_2,t_3,l_2}, where $t_2 \in U_2$ and $t_3 \in U_3$, one obtains that U_p consists of two connected components U_{p1} and U_{p2} and that for $i \in \{1,2\}$ the map

$$\kappa : U_{pi} \to U_1 \times U_2 \times U_3 : q \mapsto (\pi(q,l_1), \pi(q,l_2), \pi(q,l_3))$$

is a homeomorphism. In particular, this shows that U_p is a neighbourhood of p in $\mathcal{P}_1 \setminus \Sigma_1$.

An analogous argument shows that if p is a point such that ψ_p reverses orientation, then there is a neighbourhood U_p of p in $\mathcal{P}_1 \setminus \Sigma_1$ such that ψ_q is orientation reversing for every $q \in U_p$. This in turn implies that if p and q are two points of $\mathcal{P}_1 \setminus \Sigma_1$ such that ψ_p is orientation preserving while ψ_q is orientation reversing, then p and q are contained in different connected components of $\mathcal{P}_1 \setminus \Sigma_1$. However, the space $\mathcal{P}_1 \setminus \Sigma_1$ is connected. To see this let p and q denote two points in $\mathcal{P}_1 \setminus \Sigma_1$. First assume that $p^\perp \cap \Sigma_1 \neq q^\perp \cap \Sigma_1$. Then there exists a line l through p such that $\pi(q,l) \notin \Sigma_1$. The punctured lines $l \setminus \Sigma_1$ and $\lambda(q,l) \setminus \Sigma_1$ are homeomorphic to $\mathbb{R}$, hence connected. Moreover, the intersection $(l \setminus \Sigma_1) \cap (\lambda(q,l) \setminus \Sigma_1)$ contains the point $\pi(q,l)$, hence $(l \setminus \Sigma_1) \cup (\lambda(q,l) \setminus \Sigma_1)$ is a connected set containing p and q. If $p^\perp \cap \Sigma_1 = q^\perp \cap \Sigma_1$ choose a point $r \in q^\perp \setminus \Sigma_1$ different from q. Then $p^\perp \cap \Sigma_1 \neq r^\perp \cap \Sigma_1$, thus, by the above, the points r and p are contained in a common connected component of $\mathcal{P}_1 \setminus \Sigma_1$. But q and r are also in a common connected component, they can be joined by $(r \vee q) \setminus \Sigma_1$ which is homeomorphic to $\mathbb{R}$. Thus q and p are in a common connected component of $\mathcal{P}_1 \setminus \Sigma_1$.

With the above, connectedness of $\mathcal{P}_1 \setminus \Sigma_1$ implies that either all points in $\mathcal{P}_1 \setminus \Sigma_1$ induce orientation preserving maps or all points in $\mathcal{P}_1 \setminus \Sigma_1$ induce orientation reversing maps.

Therefore we may orient Λ_{1+} and Λ_{1-} in such a way that all induced maps ψ_p are orientation preserving. Likewise for Λ_{2+} and Λ_{2-}. That is, every point $p \in \mathcal{P}_2 \setminus \Sigma_2$ induces a homeomorphism ψ_p from Λ_{2+} to Λ_{2-} and we orient the spaces Λ_{2+} and Λ_{2-} such that these homeomorphisms are orientation preserving.

We now have a handy tool to determine whether or not the centre $s_1^\perp \cap s_2^\perp \cap s_3^\perp$ of a triad $\{s_1, s_2, s_3\}$ of K_i is empty or not. Namely, for $j \in \{1,2,3\}$ let l_{j+} and l_{j-} denote the unique line in Λ_{i+} or Λ_{i-}, respectively, that is incident with s_j. Then the map $\{l_{1+}, l_{2+}, l_{3+}\} \to \{l_{1-}, l_{2-}, l_{3-}\} : l_{j+} \mapsto l_{j-}$ can either be extended to a homeomorphism from Λ_{i+} to Λ_{i-} preserving orientation or it can be extended to a homeomorphism from Λ_{i+} to Λ_{i-} reversing orientation. In the first case we say $\{s_1, s_2, s_3\}$ is a positive triad, in the latter case we say $\{s_1, s_2, s_3\}$ is a negative triad. We claim that with orientation fixed as above the centre of a triad is not empty if and only if the triad is positive. To prove this claim consider the square projection $\mu_{s_1,s_2,l_3} : s_1^\perp \cap s_2^\perp \to l_3$. Let p denote any point in $s_1^\perp \cap s_2^\perp$. Let $t := \mu_{s_1,s_2,l_3}(p) = \pi(p, l_3)$. Then $\{s_1, s_2, t\}$ is a positive triad, since ψ_p is a orientation preserving extension. The centre $s_1^\perp \cap s_2^\perp \cap t^\perp$ contains p, hence it is not empty. Thus, for each point $u \in l_3 \setminus \{\pi(s_1, l_3), \pi(s_2, l_3)\}$ the centre $s_1^\perp \cap s_2^\perp \cap u^\perp$ is not empty and the triad $\{s_1, s_2, u\}$ is positive. Analogous arguments show that for each point u contained in the connected component of $l_3 \setminus \{\pi(s_1, l_3), \pi(s_2, l_3)\}$ not containing t the centre $s_1^\perp \cap s_2^\perp \cap u^\perp$ is empty and the triad $\{s_1, s_2, u\}$ is negative. Thus, if $\{s_1, s_2, s_3\}$ is a positive triad, then s_3 is in the same connected component of $l_3 \setminus \{\pi(s_1, l_3), \pi(s_2, l_3)\}$ as t, hence the centre $s_1^\perp \cap s_2^\perp \cap s_3^\perp$ is not empty. Conversely, if $\{s_1, s_2, s_3\}$ is a negative triad, then s_3 is in the connected component of $l_3 \setminus \{\pi(s_1, l_3), \pi(s_2, l_3)\}$ not containing t, hence the centre $s_1^\perp \cap s_2^\perp \cap s_3^\perp$ is empty.

Now we come to the construction of the map φ from the subquadrangle $\mathcal{V}_1$ fixed by τ_1 to the subquadrangle $\mathcal{V}_2$ fixed by τ_2. Take any collineation from $\mathcal{V}_1$ to $\mathcal{V}_2$. Without loss of generality we may assume that φ maps Λ_{1+} to Λ_{2+}, and thus also Λ_{1-} to Λ_{2-}. Then $\left|s_1^\perp \cap s_2^\perp \cap s_3^\perp\right| = 2 - |\varphi(s_1)^\perp \cap \varphi(s_2)^\perp \cap \varphi(s_3)^\perp|$ for any triad s_1, s_2, s_3 of $\mathcal{V}_1$ if and only if φ maps positive triads to negative triads and vice versa. This holds precisely if one of the restrictions of φ to Λ_{1+} or Λ_{1-} is orientation reversing, while the other is orientation preserving. However, such a map can always be constructed. ⚘

Theorem 5.30 and Theorem 5.31 immediately imply the next result.

(5.32) Corollary: *Suppose* $\mathfrak{GQ}_1 = (\mathcal{P}_1, \mathcal{L}_1)$ *and* $\mathfrak{GQ}_2 = (\mathcal{P}_2, \mathcal{L}_2)$ *are two antiregular compact quadrangles with parameter 1 and suppose* τ_i *are involutory collineations on* $\mathfrak{GQ}_i$ *such that* $\mathcal{V}_i := (\Sigma_i, \Lambda_i, \in)$, *where* Σ_i *denotes the fixed point set of* τ_i *and* Λ_i *the fixed line set of* τ_i, *are compact quadrangles with parameters* $(1,0)$. *Then there exists an isomorphism* $\varphi : \mathcal{V}_1 \to \mathcal{V}_2$ *such that* $\mathfrak{Mi}(\mathfrak{GQ}_1, \mathfrak{GQ}_2, \tau_1, \tau_2, \varphi)$ *is a flat Minkowski plane.* ⚘

Remarks: The above considerations imply that the two connected components of the circle space of a flat Minkowski plane $\mathfrak{Mi}$ are, geometrically, rather independent. That is, we can exchange one of the connected components of the space of circles by another set of circles, still obtaining a flat Minkowski plane. This can be done by constructing the two quadrangles $\mathfrak{GQ}_1(\mathfrak{Mi})$ and $\mathfrak{GQ}_2(\mathfrak{Mi})$ and then exchange one of these quadrangles by another, possibly quite different compact antiregular quadrangle $\mathfrak{GQ}_2'$ with parameter 1 featuring an involution τ_2' that fixes a compact subquadrangle with parameters $(1,0)$. Then, by Corollary 5.32, this leads again to a flat Minkowski plane. Most likely with this construction flat Minkowski planes with small automorphism groups can be constructed. But this is not the subject of this book.

The generalised quadrangles in the above corollary need not be different quadrangles. Thus, one can construct a flat Minkowski plane from just one compact quadrangle with parameter 1. More precisely, from Corollary 5.32 we immediately obtain the following result:

(5.33) Corollary: *Suppose* $\mathfrak{GQ} = (\mathcal{P}, \mathcal{L})$ *is a compact antiregular quadrangle with parameter 1 and suppose* τ *is an involutory collineation on* $\mathfrak{GQ}$ *such that* $\mathcal{V} := (\Sigma, \Lambda, \in)$, *where* Σ *denotes the fixed point set of* τ *and* Λ *the fixed line set of* τ, *is a compact quadrangle with parameters* $(1,0)$. *Then there exists an automorphism* $\varphi : \mathcal{V} \to \mathcal{V}$ *such that* $\mathfrak{Mi}(\mathfrak{GQ}, \mathfrak{GQ}, \tau, \tau, \varphi)$ *is a flat Minkowski plane.* ⚘

We now come to the parameter 2 case. The first step is to show that the quadrangle constructed from a locally compact Minkowski plane with parameter 2 is centric antiregular.

(5.34) Lemma: *If* $\mathfrak{Mi}$ *is a locally compact Minkowski plane with parameter 2, then the quadrangle* $\mathfrak{GQ}(\mathfrak{Mi})$ *is centric antiregular.*

Proof: By Theorem 5.26 the lifted Lie geometry $\mathfrak{GQ}(\mathfrak{Mi})$ is a compact quadrangle with parameter 2. Thus we only have to show that this quadrangle is antiregular (Corollary 2.10). By Lemma 2.3 we know that for a compact quadrangle with parameter 2 the set $p^\perp$ is homeomorphic to the one-point-compactification of the tangent bundle of $\mathbb{S}_2$ for every point p and the set $g^\perp$ is homeomorphic to the complex projective plane for every line g, or vice versa.

The point space $\mathcal{S}$ of the Minkowski plane and the plus- and minus-parallel classes of the Minkowski plane form a compact subquadrangle with parameters $(2,0)$. Let l_1 and l_2 denote a pair of non-intersecting lines in this subquadrangle. So we may assume that l_1 and l_2 correspond to two plus-parallel classes of the Minkowski plane $\mathfrak{Mi}$. Then $l_1^\perp \cap l_2^\perp$ consists precisely of the lines corresponding to minus-parallel classes, i.e., $l_1^\perp \cap l_2^\perp = \mathcal{G}_-$. Let p denote a point not contained in the subquadrangle. Then p corresponds to a cycle k of the Minkowski plane and $p^\perp \cap \mathcal{S}$ consists precisely of the points of the Minkowski plane incident with k. Now consider the map $\mu : l_1^\perp \cap l_2^\perp \to \mathcal{L}_p$. This map is a square projection in the dual quadrangle. If g is a line in $l_1^\perp \cap l_2^\perp$, then there is a point s of the Minkowski plane such that $g = \Gamma_-(s)$ and $\mu(g) = \widetilde{B}(\gamma_-(s,k),k)$. In particular μ is injective. This implies that the twisting number of the dual quadrangle equals 1, hence the twisting number of the quadrangle $\mathfrak{GQ}(\mathfrak{Mi})$ equals 2 by Theorem 2.15. By the same theorem the quadrangle is antiregular. ⚘

As in the flat case we can characterise those compact antiregular quadrangles with parameter 2 that can be obtained from Minkowski planes. Again, the twofold covering $c : \widetilde{\mathcal{K}} \to \mathcal{K}$ induces a continuous fixed point free involution $\tau_{\mathfrak{Mi}}$ on $\widetilde{\mathcal{K}}$ such that $c(p) = c(\tau_{\mathfrak{Mi}}(p))$ for every cycle $p \in \widetilde{\mathcal{K}}$. We extend this involution by the identity on $\mathcal{S}$ to the whole point space of the lifted Lie geometry. The extended map $\tau_{\mathfrak{Mi}}$ acts also as an involution on the line space of the lifted Lie geometry leaving precisely the lines in $\mathcal{G}_+ \cup \mathcal{G}_-$ fixed. Thus we obtain the following result:

(5.35) Theorem: *Suppose* $\mathfrak{GQ} = \mathfrak{GQ}(\mathfrak{Mi})$ *is the lifted Lie geometry of a locally compact Minkowski plane* $\mathfrak{Mi}$ *with parameter 2. Then* $\mathfrak{GQ}$ *is a compact centric antiregular quadrangle with parameter 2 and there is a continuous involutory collineation* $\tau_{\mathfrak{Mi}}$ *such that the space of fixed points and the space of fixed lines form a compact subquadrangle* $\mathcal{V}$ *with parameters* (2,0). ☺

As before, the construction can be reversed. The reversed construction works also for the non-topological case.

(5.36) Theorem: *Suppose* $\mathfrak{GQ} = (\mathcal{P}, \mathcal{L})$ *is a thick centric antiregular quadrangle with an involutory collineation* τ *such that the space* Σ *of fixed points and the space* Λ *of fixed lines form a point thin subquadrangle* $\mathcal{V}$. *Then*

$$\mathfrak{Mi}(\mathfrak{GQ}, \tau) := (\Sigma, \{\Sigma \cap q^\perp \mid q \in \mathcal{P} \setminus \Sigma\}, \Lambda, \in)$$

is a Minkowski plane.

Proof: The theorem is proved much along the same lines as Theorem 5.29.

Let $K := \{\Sigma \cap q^\perp \mid q \in \mathcal{P} \setminus \Sigma\}$ and let $\kappa : \mathcal{P} \setminus \Sigma \to K : q \mapsto q^\perp \cap \Sigma$. The set Λ splits into two parts Λ_+ and Λ_- such that every line in Λ_+ meets every line in Λ_- but any two lines in Λ_+ and any two lines in Λ_- do not intersect. In anticipation of the claimed result we will call the elements of K circles, the elements of Λ_+ plus-parallel classes and the elements of Λ_- minus-parallel classes.

First we show that any three mutually non-collinear points in Σ define a unique circle in K. So suppose $s_1, s_2, s_3 \in \Sigma$ form a triad. Then, by centric antiregularity, the centre $s_1^\perp \cap s_2^\perp \cap s_3^\perp$ contains precisely two points p_1 and p_2. Since every point in Σ is contained only in two lines of Λ we can deduce that neither p_1 nor p_2 is in Σ. Thus $\tau(p_1) = p_2$. Hence $\{s_1, s_2, s_3\} \in \kappa(p_1)$ and $\kappa(p_1) = \Sigma \cap p_1^\perp = \tau(\Sigma \cap p_1^\perp) = \tau(\Sigma) \cap \tau(p_1^\perp) = \Sigma \cap \tau(p_1)^\perp = \kappa(p_2) \in K$ is the unique element $\alpha(s_1, s_2, s_3)$ of K incident with s_1, s_2, s_3. We also obtain that the map κ is two to one.

Next we want to define and discuss the map β which is supposed to be the touching map as in the definition of a Minkowski plane. Suppose $s, t \in \Sigma$, $k \in K$, such that $s \in k$ and such that t is not collinear with s. Take $p \in \kappa^{-1}(k)$ and let $g \in \mathcal{L}$ denote the line joining s to p. Since t is not collinear with s, the point $\pi(t, g)$ is not contained in Σ. Moreover, $\tau(g)$ is the line joining s to $\tau(p)$ and $\beta(s, k, t) := \kappa(\pi(t, g)) = \kappa(\pi(t, \tau(g)) \in K$ is the unique element of K incident with t and touching k in s.

Our next step is to investigate parallel projections. Suppose $s \in \Sigma$ and $k \in K$. Let l_+ denote the unique line in Λ_+ incident with s and let l_- denote the unique line in Λ_- incident with s. Let $p \in \kappa^{-1}(k)$. Then $\gamma_+(s, k) := \pi(p, l_+) = k \cap l_+$ is the unique point on k plus-parallel to s and $\gamma_-(s, k) := \pi(p, \lambda(p, l_-) = k \cap l_-$ is the unique point on k minus-parallel to s.

We now discuss the map δ mentioned in the axioms for a Minkowski plane. Let l_- denote the line in Λ_- incident with t. Then $\delta(s, t) := \pi(s, l_-)$ is the unique point that is plus-parallel to s and minus-parallel to t.

Thickness of the quadrangle $\mathfrak{GQ}$ implies that the derived structure is not degenerate. Thus $\mathfrak{Mi}(\mathfrak{GQ}, \tau)$ is a Minkowski plane. ♋

Restricting our attention to the topological case we obtain the next result.

(5.37) Theorem: *Suppose $\mathfrak{GQ} = \mathfrak{GQ}(\mathfrak{Mi})$ is the lifted Lie geometry of a locally compact Minkowski plane $\mathfrak{Mi}$ with parameter 2. Then $\mathfrak{GQ}$ is a compact centric antiregular quadrangle with parameter 2 and there is a continuous involutory collineation $\tau_{\mathfrak{Mi}}$ such that the space of fixed points and the space of fixed lines form a compact subquadrangle $\mathcal{V}$ with parameters $(2,0)$.*

Conversely, if $\mathfrak{GQ} = (\mathcal{P}, \mathcal{L})$ is a compact antiregular quadrangle with parameter 2 and a map τ as above, then the derived structure

$$\mathfrak{Mi}(\mathfrak{GQ}, \tau) = \left(\Sigma, \{\Sigma \cap q^{\perp} \mid q \in \mathcal{P} \setminus \Sigma\}, \Lambda_+ \cup \Lambda_-, \in\right),$$

where Σ denotes the space of points fixed by τ and where Λ_+ and Λ_- denote the two connected components of the space of lines fixed by τ, is a locally compact Minkowski $\mathfrak{Mi}$ plane with parameter 2.

Proof: The first part of the theorem follows from Theorem 5.35. In Theorem 5.36 it was shown that the derived structure is a Minkowski plane. Thus we only have to prove that this Minkowski plane has parameter 2. However, now we have to make sure directly that the geometric operations of the Minkowski plane obtained are continuous. This is accomplished having a closer look at the construction of these operations.

As before let $K := \{\Sigma \cap q^\perp \,|\, q \in \mathcal{P} \setminus \Sigma\}$ and let $\kappa : \mathcal{P} \setminus \Sigma \to K : q \mapsto q^\perp \cap \Sigma$. Parallel classes, i.e., the elements of $\Lambda_+ \cup \Lambda_-$ correspond to lines of the quadrangle fixed by τ, hence they are homeomorphic to $\mathbb{S}_2$. The point space Σ is the product of two lines, hence homeomorphic to $\mathbb{S}_2 \times \mathbb{S}_2$. Circles are homeomorphic to traces of the quadrangle, hence to $\mathbb{S}_2$.

We now check that the map κ is continuous. By [11]: 2.12 we know that if a sequence of points $p_i \in \mathcal{P}$, $i \in \mathbb{N}$, converges to some point $p \in \mathcal{P}$, then $p_i^\perp$ converges to $p^\perp$ with respect to the topology induced by the Hausdorff metric. Thus $\kappa(p_i)$, $i \in \mathbb{N}$, converges to a subset of $\kappa(p)$. On the other hand, if $s \in \kappa(p)$, let $g \in \Lambda_+$ be the line incident with s. Then for every $i \in \mathbb{N}$ the intersection $g \cap \kappa(p_i)$ contains precisely one point s_i, namely $s_i = \pi(p_i, g)$. By continuity of the map π, the sequence $s_i \in \kappa(p_i)$ converges to s. This shows that the sequence $\kappa(p_i)$, $i \in \mathbb{N}$, converges to $\kappa(p)$.

The proof of Theorem 5.36 shows that the map κ is two to one. By [43] the centre $s_1^\perp \cap s_2^\perp \cap s_3^\perp$ depends continuously on s_1, s_2 and s_3, thus the map κ is even a twofold covering, which in turn implies that the map α is continuous.

Suppose $s, t \in \Sigma$, $k \in K$, such that $s \in k$ and such that t is not collinear with s. Then $\beta(s, k, t) = \kappa(\pi(t, g)) = \kappa(\pi(t, \tau(g))$ where g is the line joining s to $p \in \kappa^{-1}(k)$. Hence, since the map κ is a twofold covering and since the map π is continuous, the map β is continuous as well.

Our next step is to show that parallel projections are continuous. Suppose $s \in \Sigma$ and $k \in K$. Choose $g_+ \in \Lambda_+$ and $g_- \in \Lambda_-$ such that s is neither incident with g_+ nor with g_- and choose $p \in \kappa^{-1}(k)$. Then $\gamma_+(s, k) = \pi(p, \lambda(s, g_-))$ and $\gamma_-(s, k) = \pi(p, \lambda(s, g_+))$. Since the lines g_+ and g_- are closed in Σ, there is a neighbourhood of s in Σ containing no point of g_+ or g_- and therefore continuity of π and λ and the fact that the map κ is a twofold covering imply that the maps γ_+ and γ_- are continuous at $(s, k) \in \Sigma \times K$.

We now discuss the map δ mentioned in the axioms for a Minkowski plane. Suppose $s, t \in \Sigma$ are two non-collinear points. By [11]: 2.13 the pairs of non-collinear points are open in $\mathcal{P}$. Therefore there are neighbourhoods U_s and U_t of s and t in Σ respectively, such that no point in U_s is collinear with a point in U_t. Let l denote the line in Λ_- incident with s. Then $\delta(u, v) = \pi(v, \lambda(u, l))$. Hence δ is continuous on $U_s \times U_t$. Since t and s are arbitrary non-parallel points, the map δ is continuous.

Thus we now know that all maps appearing in the definition of a Minkowski plane are continuous. What remains to be shown is that the intersection of two circles depends continuously on the two circles. That is, we have to show that if (k_i, l_i), $i \in \mathbb{N}$, is a sequence of pairs of non-tangent circles converging to a pair (k, l) of non-tangent circles, then $k_i \cap l_i$ converges to $k \cap l$ with respect to the Hausdorff metric.

Let $\{a_i, b_i\} = \kappa^{-1}(k_i)$, and let $c_i \in \kappa^{-1}(l_i)$. Since the map κ is a twofold covering, we may assume that the sequence (a_i, b_i, c_i), $i \in \mathbb{N}$, converges to (a, b, c) where $\{a, b\} = \kappa^{-1}(k)$ and $c \in \kappa^{-1}(l)$. Then $k_i = a_i^\perp \cap b_i^\perp$, $i \in \mathbb{N}$, and $k = a^\perp \cap b^\perp$. Thus $\lim_{i\to\infty} k_i \cap l_i = \lim_{i\to\infty} a_i^\perp \cap b_i^\perp \cap c_i^\perp = a^\perp \cap b^\perp \cap c^\perp = k \cap l$ by the fact that the centre $s_1^\perp \cap s_2^\perp \cap s_3^\perp$ depends continuously on s_1, s_2 and s_3 ([43]). ⚘

As in the case of the other circle geometries we now briefly look at the group of automorphisms. Again, if α is an element of a group G let $\mathrm{Cs}_\alpha(G)$ denote the subgroup of elements of G commuting with α. With the same arguments as in the case of flat Möbius planes (Corollary 4.17 and Corollary 4.18) we obtain the following results:

(5.38) Corollary: *Suppose $\mathfrak{G}\mathfrak{Q}$ is a compact antiregular quadrangle with parameter 2 and suppose τ is an involution acting on $\mathfrak{G}\mathfrak{Q}$ as specified in Theorem 5.37. Then*

$$\mathrm{Aut}(\mathfrak{Mi}(\mathfrak{G}\mathfrak{Q}, \tau)) = \mathrm{Cs}_\tau(\mathrm{Aut}(\mathfrak{G}\mathfrak{Q}))/\langle\tau\rangle.$$

Conversely, if $\mathfrak{Mi}$ is a locally compact Minkowski plane with parameter 2, then

$$\mathrm{Aut}(\mathfrak{Mi}) = \mathrm{Cs}_{\tau_{\mathfrak{Mi}}}(\mathrm{Aut}(\mathfrak{G}\mathfrak{Q}(\mathfrak{Mi})))/\langle\tau\rangle.$$

⚘

(5.39) Corollary: *Suppose $\mathfrak{G}\mathfrak{Q}$ is a compact antiregular quadrangle with parameter 2 and suppose τ_1 is an involution acting on $\mathfrak{G}\mathfrak{Q}$ as specified in Theorem 5.37. Suppose τ_2 is another involution on $\mathfrak{G}\mathfrak{Q}$. Then τ_2 acts on $\mathfrak{G}\mathfrak{Q}$ as specified in Theorem 5.37 and $\mathfrak{Mi}(\mathfrak{G}\mathfrak{Q}, \tau_1) \cong \mathfrak{Mi}(\mathfrak{G}\mathfrak{Q}, \tau_2)$ if and only if there is an automorphism ψ of $\mathfrak{G}\mathfrak{Q}$ such that $\tau_2 = \psi\tau_1\psi^{-1}$.* ⚘

In the general, i.e., non-topological case we obtain:

(5.40) Corollary: *Suppose $\mathfrak{G}\mathfrak{Q}$ is a thick centric antiregular quadrangle. If τ is an involution acting on $\mathfrak{G}\mathfrak{Q}$ as specified in Theorem 5.36, then*

$$\mathrm{Aut}(\mathfrak{Mi}(\mathfrak{G}\mathfrak{Q}, \tau)) = \mathrm{Cs}_\tau(\mathrm{Aut}(\mathfrak{G}\mathfrak{Q}))/\langle\tau\rangle.$$

Suppose τ_2 is another involution on $\mathfrak{GQ}$. Then τ_2 acts on $\mathfrak{GQ}$ as specified in Theorem 5.36 and $\mathfrak{Mi}(\mathfrak{GQ}, \tau_1) \cong \mathfrak{Mi}(\mathfrak{GQ}, \tau_2)$ if and only if there is an automorphism ψ of $\mathfrak{GQ}$ such that $\tau_2 = \psi\tau_1\psi^{-1}$. ۞

Remark: For flat Minkowski planes and pairs of non-centric quadrangles the situation is not so obvious. The group of automorphisms of the Minkowski plane $\mathfrak{Mi}$ can be a proper subgroup of the groups $\mathrm{Cs}_{\tau_i}(\mathrm{Aut}(\mathfrak{GQ}_i))$. The difference depends very much on the map φ. If the map $\varphi : \mathcal{V}_1 \to \mathcal{V}_2$ can be extended to a collineation from $\mathfrak{GQ}_1$ to $\mathfrak{GQ}_2$, then the situation is different. More precisely, $\mathrm{Aut}(\mathfrak{GQ}_1) \cong \mathrm{Aut}(\mathfrak{GQ}_2)$ and $\mathrm{Cs}_{\tau_i}(\mathrm{Aut}(\mathfrak{GQ}_i))/\langle\tau_i\rangle$ is a subgroup of index two in $\mathrm{Aut}(\mathfrak{Mi})$. But, in this book, we will no longer linger on this line of thought.

6 Relationships of circle planes

Introduction

In the previous chapters we investigated the connection between circle planes and generalised quadrangles. We saw that given any finite-dimensional connected circle plane one can construct a compact antiregular quadrangle or a pair of compact antiregular quadrangles. Conversely, given an antiregular quadrangle, one can construct a Laguerre plane for every point of this quadrangle. If, in addition, the quadrangle features an involution with certain properties, one can also construct a Minkowski plane or a Möbius plane, depending on the properties of the involution.

We can look at these results also from a slightly different point of view. First we define certain sets of geometries. Let $\mathcal{GQ}_t$, with $t \in \{1,2\}$, denote the set of compact quadrangle with parameter t. Related to this set are the following sets of geometries. Let $\mathcal{GQP}_t := \{(\mathfrak{GQ}, p) \mid \mathfrak{GQ} = (\mathcal{P}, \mathcal{L}) \in \mathcal{GQ}_t \quad \text{and} \quad p \in \mathcal{P}\}$. Let $\mathcal{GQT}_1$ denote the set of pairs $(\mathfrak{GQ}, \tau)$, where $\mathfrak{GQ} \in \mathcal{GQ}_1$ and where τ is an involutory collineation on $\mathfrak{GQ}$ meeting the requirements of Proposition 4.16. Let $\mathcal{GQS}_2$ denote the set of pairs $(\mathfrak{GQ}, \tau)$, where $\mathfrak{GQ} \in \mathcal{GQ}_2$ and where τ is an involutory collineation on $\mathfrak{GQ}$ meeting the requirements of Theorem 5.37. Finally, let $\mathcal{GQS}_1$ denote the set of 5-tuples $(\mathfrak{GQ}_1, \mathfrak{GQ}_2, \tau_1, \tau_2, \varphi)$, where $\mathfrak{GQ}_i \in \mathcal{GQ}_1$, where τ_i are involutory collineations on $\mathfrak{GQ}_i$ meeting the requirements of Theorem 5.30 and where φ i fixed by τ_1 to the subquadrangle fixed by τ_2 meeting the requirements of Theorem 5.30. Projection onto the first coordinate yields natural maps from $\mathcal{GQP}_t$, $\mathcal{GQT}_t$ and $\mathcal{GQS}_t$ to $\mathcal{GQ}_t$. We also consider sets of circle planes. Let $\mathcal{LA}_t$ denote the set of locally compact Laguerre planes with parameter t, let $\mathcal{MO}_1$ denote the set of flat Möbius planes and let $\mathcal{MI}_t$ denote the set of locally compact Minkowski planes with parameter t.

In previous chapters we constructed bijections between some of these sets. In Chapter 3 we constructed the bijection

$$\mathfrak{GQ}_{\mathfrak{L}} : \mathcal{LA}_t \to \mathcal{GQ}_t : \mathfrak{L} \mapsto (\mathfrak{GQ}(\mathfrak{L}), \infty_{\mathfrak{L}}),$$

where $\infty_{\mathfrak{L}}$ is the additional point specified by the construction. Its inverse is the map

$$\mathfrak{GQ}_{\mathfrak{L}}^{-1} : \mathcal{GQ}_t \to \mathcal{LA}_t : (\mathfrak{GQ}, p) \mapsto \mathfrak{L}(\mathfrak{GQ}, p).$$

In Chapter 4 we introduced the bijection

$$\mathfrak{GQ}_{\mathfrak{Mo}} : \mathcal{MO}_1 \to \mathcal{GQT}_1 : \mathfrak{Mo} \mapsto (\mathfrak{GQ}(\mathfrak{Mo}), \tau_{\mathfrak{Mo}}),$$

where $\tau_{\mathfrak{Mo}}$ is the involution on the quadrangle induced by reversing orientation of cycles. The inverse map is

$$\mathfrak{GQ}_{\mathfrak{Mo}}^{-1} : \mathcal{GQT}_1 \to \mathcal{MO}_1 : (\mathfrak{GQ}, \tau) \mapsto \mathfrak{Mo}(\mathfrak{GQ}, \tau).$$

In Chapter 5 we investigated the connection between Minkowski planes and generalised quadrangles. For the parameter 2 case we obtained the bijection

$$\mathfrak{GQ}_{\mathfrak{Mi}} : \mathcal{MI}_2 \to \mathcal{GQS}_2 : \mathfrak{Mi} \mapsto (\mathfrak{GQ}(\mathfrak{Mi}), \tau_{\mathfrak{Mi}}),$$

where $\tau_{\mathfrak{Mi}}$ is the involution on the quadrangle induced by reversing orientation of cycles. Its inverse is the map

$$\mathfrak{GQ}_{\mathfrak{Mi}}^{-1} : \mathcal{GQS}_2 \to \mathcal{MI}_2 : (\mathfrak{GQ}, \tau) \mapsto \mathfrak{Mi}(\mathfrak{GQ}, \tau).$$

For the parameter 1 case we obtained the bijection

$$\mathfrak{GQ}_{\mathfrak{Mi}} : \mathcal{MI}_1 \to \mathcal{GQS}_1 : \mathfrak{Mi} \mapsto (\mathfrak{GQ}_1(\mathfrak{Mi}), \mathfrak{GQ}_2(\mathfrak{Mi}), \tau_{1,\mathfrak{Mi}}, \tau_{2,\mathfrak{Mi}}, \varphi),$$

where $\tau_{1,\mathfrak{Mi}}$ and $\tau_{2,\mathfrak{Mi}}$ are involutions on $\mathfrak{GQ}_1(\mathfrak{Mi})$ and $\mathfrak{GQ}_2(\mathfrak{Mi})$ respectively induced by reversing orientation of cycles and where φ is a map from the set of points fixed by $\tau_{1,\mathfrak{Mi}}$ to the set of points fixed by $\tau_{2,\mathfrak{Mi}}$ specified by the construction. The inverse of $\mathfrak{GQ}_{\mathfrak{Mi}}$ is the map

$$\mathfrak{GQ}_{\mathfrak{Mi}}^{-1} : \mathcal{GQS}_1 \to \mathcal{MI}_1 : (\mathfrak{GQ}_1, \mathfrak{GQ}_2, \tau_1, \tau_2, \varphi) \mapsto \mathfrak{Mi}(\mathfrak{GQ}_1, \mathfrak{GQ}_2, \tau_1, \tau_2, \varphi).$$

We summarise these relationships in the following diagram, where $\mathcal{GQT}_t$ and $\mathcal{MO}_t$ are void for $t \neq 1$:

$$\begin{array}{ccc} & \mathcal{GQ}_t & \\ \nearrow & \uparrow & \nwarrow \\ \mathcal{GQP}_t & \mathcal{GQT}_t & \mathcal{GQS}_t \\ \uparrow \scriptstyle{\mathfrak{GQ}_{\mathfrak{L}}} & \uparrow \scriptstyle{\mathfrak{GQ}_{\mathfrak{Mo}}} & \uparrow \scriptstyle{\mathfrak{GQ}_{\mathfrak{Mi}}} \\ \mathcal{LA}_t & \mathcal{MO}_t & \mathcal{MI}_t \end{array}$$

The results of the previous chapters imply that one can define morphisms on the above sets of geometries such that one obtains categories. Then the maps $\mathfrak{GQ}_{\mathfrak{L}}$, $\mathfrak{GQ}_{\mathfrak{Mo}}$ and $\mathfrak{GQ}_{\mathfrak{Mi}}$ as well as their inverse maps become covariant functors.

Notation: To stress the close relationship between the different geometries involved, also the following notation is used. The generalised quadrangle constructed from a circle plane $\mathfrak{Cp}$ is called the **mother** of $\mathfrak{Cp}$. A Laguerre plane obtained from a generalised quadrangle $\mathfrak{GQ}$ is called a **daughter** of $\mathfrak{GQ}$ and a Möbius- or Minkowski plane obtained from a generalised quadrangle $\mathfrak{GQ}$ is called a **son** of $\mathfrak{GQ}$.

Well, what about sisters and brothers? Naturally, two circle geometries are sisters or brothers if they have the same mother. We will see in this chapter that every son has sisters. But not every daughter has a brother. However, every daughter has quite a number of sisters. In order to construct sisters and brothers we look at compositions of variations of the maps $\mathfrak{GQ}_{\mathfrak{L}}$, $\mathfrak{GQ}_{\mathfrak{Mo}}$ and $\mathfrak{GQ}_{\mathfrak{Mi}}$ as well as their inverse maps.

In the first section we look at sisters of Laguerre planes. In other terms, we construct Laguerre planes from other Laguerre planes.

In the second section we look at sisters of Möbius planes. That is, we construct flat Laguerre planes from flat Möbius planes. For the classical case this construction is rather old. In fact it is the original construction of a Laguerre plane in terms of oriented circles and oriented lines ([23]). For the topological case this construction was already introduced by H. Groh in [15]. Although his construction is more or less the same as the one presented here, his proofs are different. Moreover, the construction presented in this book is more general. We also give a characterisation of the Laguerre planes obtained from Möbius planes, i.e., of the Laguerre planes that have a Möbius plane as a brother.

In the third section we look at sisters of Minkowski planes. In other words, we construct Laguerre planes from Minkowski planes. This construction has no known classical counterpart. Most likely, this is due to the fact that in Minkowski planes oriented circles are not as easy to visualise as in Möbius planes. Again, we will characterise all Laguerre planes one can obtain from Minkowski planes, i.e., all Laguerre planes that have a Minkowski plane as a brother.

It is also very tempting to look at brothers of a Möbius or Minkowski plane. However, within circle planes brotherhoods cannot be described easily, hence we will not investigate them in this book. There is also the question whether or not there are brotherhoods of non-classical Minkowski planes. That is, is there a non-classical compact antiregular quadrangle with parameter 1 and two involutions τ_1 and τ_2, such that τ_1 meets the requirements of Proposition 4.16 and τ_2 meets the requirements of Corollary 5.33? No attempt has yet been made to answer this question, but my belief is that there are such quadrangles.

6.1 Sisters of Laguerre planes

Suppose $\mathfrak{L} = (\mathcal{S}, \mathcal{K})$ is a finite-dimensional connected Laguerre plane. For every element $p \in \mathcal{S} \cup \mathcal{K} \cup \{\infty_{\mathfrak{L}}\}$, where $\infty_{\mathfrak{L}}$ denotes the additional point specified by the construction of $\mathfrak{G}\mathfrak{Q}(\mathfrak{L})$, we define the **sister $\mathfrak{L}(\mathfrak{L}, p)$ of $\mathfrak{L}$ at p** as follows:

$$\mathfrak{L}(\mathfrak{L}, p) := \mathfrak{L}(\mathfrak{G}\mathfrak{Q}(\mathfrak{L}), p).$$

From Theorem 3.2, Theorem 3.13 and Theorem 3.14 we immediately obtain the following result:

(6.1) Theorem: *Suppose $\mathfrak{L} = (\mathcal{S}, \mathcal{K})$ is a locally compact Laguerre plane with parameter t. Then, for every $p \in \mathcal{S} \cup \mathcal{K} \cup \{\infty_{\mathfrak{L}}\}$, the sister $\mathfrak{L}(\mathfrak{L}, p)$ at p is again a locally compact Laguerre plane with parameter t.*

Moreover, $\mathfrak{L}(\mathfrak{L}, \infty_{\mathfrak{L}}) = \mathfrak{L}$. The sister $\mathfrak{L}(\mathfrak{L}, p)$ is (topologically) isomorphic to $\mathfrak{L}$ if and only if there is a (topological) automorphism of $\mathfrak{G}\mathfrak{Q}(\mathfrak{L})$ mapping $\infty_{\mathfrak{L}}$ to p. ۞

The above definition of the sister of a Laguerre plane is rather neat and compact but it uses a quadrangle as an intermediate step. One can however describe a sister of a Laguerre plane also entirely within the Laguerre plane.

If $p \in \mathcal{S}$, then the points of $\mathfrak{L}(\mathfrak{L}, p)$ are all points different from p but parallel to p plus all circles incident with p plus an additional point $\infty_{\mathfrak{L}}$. The circles of $\mathfrak{L}(\mathfrak{L}, p)$ are all points of $\mathfrak{L}$ not parallel to p and all circles not incident with p. A flag of $\mathfrak{L}(\mathfrak{L}, p)$ consists either of a pair of parallel points or of a flag of $\mathfrak{L}$ or of a pair of tangent circles or of a pair $(\infty_{\mathfrak{L}}, q)$, where q is a point not parallel to p.

If $p \in \mathcal{K}$, then the points of $\mathfrak{L}(\mathfrak{L}, p)$ are all points incident with p plus all circles different from p but tangent to p. The circles of $\mathfrak{L}(\mathfrak{L}, p)$ are all points not incident with p, the additional point $\infty_{\mathfrak{L}}$ plus all circles not tangent to p. A flag of $\mathfrak{L}(\mathfrak{L}, p)$ consists either of pair of parallel points, or of a flag of $\mathfrak{L}$ or of a pair of tangent circles or of a pair $(q, \infty_{\mathfrak{L}})$, where q is a point incident with p.

Sisterhoods of Laguerre planes might be used to construct "new" Laguerre planes. Sisterhoods might also be used to discover new connections between known Laguerre planes. The automorphism group of the quadrangle constructed from the real or complex classical Laguerre plane is transitive on the set of points. Thus the sister of a classical Laguerre plane is again a classical Laguerre plane. The situation is different for non-classical Laguerre planes. The most striking relationship between different Laguerre planes yet discovered occurs if one considers the Laguerre planes of "shear type" and "translation type" introduced in [27] and [28]. Computation reveals that the sister of a Laguerre plane of shear type at a point specified by the action of the group is a Laguerre plane of translation type and vice versa. This result will be published in more detail in [44].

6.2 Sisters of Möbius planes

Now start with a flat Möbius plane $\mathfrak{Mo}$ rather than with a flat Laguerre plane. By Theorem 4.11 the quadrangle $\mathfrak{GQ} = (\mathcal{P}, \mathcal{L}) := \mathfrak{GQ}(\mathfrak{Mo})$ is antiregular and compact with parameter 1. Thus, for every point p of this quadrangle the derived structure $\mathfrak{L}(\mathfrak{GQ}, p)$ yields a flat Laguerre plane. Moreover, if p corresponds to a point of the Möbius plane rather than to a cycle, then the involution $\tau_{\mathfrak{Mo}}$ acting on the quadrangle which is induced by reversing orientation of cycles acts also on the derived Laguerre plane, since $\tau_{\mathfrak{Mo}}$ fixes p. To avoid confusion we will denote the induced involution on the Laguerre plane by $\boldsymbol{\vartheta}_{\mathfrak{Mo}}$. The properties of $\tau_{\mathfrak{Mo}}$ listed in Proposition 4.16 will induce certain properties of $\vartheta_{\mathfrak{Mo}}$. Conversely, if we are given a flat Laguerre plane with an involution ϑ with certain properties, we can construct a flat Möbius plane by reversing the above construction. More precisely, we have the following theorem:

(6.2) Theorem: *Suppose p is a point or a cycle of a flat Möbius plane $\mathfrak{Mo}$. Then*

$$\mathfrak{L}(\mathfrak{Mo}, p) := \mathfrak{L}(\mathfrak{GQ}(\mathfrak{Mo}), p)$$

is a flat Laguerre plane, called the sister of $\mathfrak{Mo}$ at p.

Suppose p is a point of a flat Möbius plane. Then the involution $\vartheta_{\mathfrak{Mo}}$ induced on $\mathfrak{L}(\mathfrak{Mo}, p)$ by reversing orientation of cycles acts without fixed points and there are two intersecting circles of $\mathfrak{L}(\mathfrak{Mo}, p)$ fixed by $\vartheta_{\mathfrak{Mo}}$.

Conversely, if ϑ is an involution on a flat Laguerre plane $\mathfrak{L}$ with the above properties, then

$$\mathfrak{Mo}(\mathfrak{L}, \vartheta) := \mathfrak{Mo}(\mathfrak{GQ}(\mathfrak{L}), \tau),$$

where τ is the involution on $\mathfrak{GQ}(\mathfrak{L})$ induced by ϑ, is a flat Möbius plane, called the **brother** *of $\mathfrak{L}$.*

The above constructions are inverse to each other: If ϑ is an involution on a flat Laguerre plane $\mathfrak{L}$ with the above properties, then

$$\mathfrak{L}(\mathfrak{Mo}(\mathfrak{L}, \vartheta), \infty_{\mathfrak{L}}) = \mathfrak{L} \qquad \text{and} \qquad \vartheta_{\mathfrak{Mo}(\mathfrak{L},\vartheta)} = \vartheta.$$

Conversely,

$$\mathfrak{Mo}(\mathfrak{L}(\mathfrak{Mo},p),\vartheta_{\mathfrak{Mo}}) = \mathfrak{Mo} \qquad \text{and} \qquad \infty_{\mathfrak{L}(\mathfrak{Mo},p)} = p$$

for every point or cycle p of a flat Möbius plane $\mathfrak{Mo}$.

Proof: It follows readily from Theorem 4.15 and Theorem 3.2 that $\mathfrak{L}(\mathfrak{Mo},p)$ is a flat Laguerre plane.

Suppose p is a point of a flat Möbius plane. Let $\tau := \tau_{\mathfrak{Mo}}$ denote the involution on $\mathfrak{GQ}(\mathfrak{Mo})$ induced by reversing orientation of cycles. Then τ fixes p and therefore acts on $p^\perp\backslash\{p\}$, i.e., τ induces an involution $\vartheta := \vartheta_{\mathfrak{Mo}}$ on $\mathfrak{L}(\mathfrak{Mo},p)$. By Theorem 4.15 no lines are fixed by τ, thus no point collinear with p is fixed by τ and therefore ϑ acts without fixed points. By Proposition 4.16, there are two points s and t fixed by τ and different from p such that $p^\perp \cap s^\perp \cap t^\perp = (p^\perp \cap s^\perp) \cap (p^\perp \cap t^\perp) \neq \emptyset$. Thus, there are two intersecting circles $p^\perp \cap s^\perp$ and $p^\perp \cap t^\perp$ of $\mathfrak{L}(\mathfrak{Mo},p)$ fixed by ϑ.

Conversely, suppose $\mathfrak{L}$ is a flat Laguerre plane and ϑ an involution on $\mathfrak{L}$ without fixed points such that there are two intersecting circles k and l fixed by ϑ. Let τ denote the corresponding involution on the quadrangle $\mathfrak{GQ}(\mathfrak{L})$, let $p := \infty_{\mathfrak{L}}$ and let s and t denote the points of this quadrangle corresponding to k and l. If g is a line not incident with p, then the point $\pi(p,g)$ is moved by τ, since τ acts without fixed points on $p^\perp \setminus \{p\}$. Assume g is fixed by τ. Then $\tau(\pi(p,g)) = \pi(\tau(p),\tau(g)) = \pi(p,g)$ is fixed, contradicting the above. Therefore, no line not incident with p is fixed. If g is a line incident with p, then g is not incident with s and the point $\pi(s,g)$ which is collinear with p is not fixed by τ. Again, this implies that τ does not fix g, hence τ fixes no line. Moreover, the points p, s and t are fixed by τ and the intersection $p^\perp \cap s^\perp \cap t^\perp$ is not empty, since it contains the points of the quadrangle corresponding to the points in $k \cap l$. Therefore, by Proposition 4.16, the geometry $\mathfrak{Mo}(\mathfrak{L},\vartheta) = \mathfrak{Mo}(\mathfrak{GQ}(\mathfrak{L}),\tau)$ is a flat Möbius plane.

That the constructions are inverse to each other is obvious. ☺

Remarks: If p is a cycle of a flat Möbius plane and τ the induced involution on $\mathfrak{GQ}(\mathfrak{Mo})$, then τ induces no collineation on $\mathfrak{L}(\mathfrak{Mo},p)$, since $p^\perp$ is not invariant under τ. Therefore, the flat Laguerre planes obtained from a flat Möbius plane via a cycle cannot be characterised internally by some involutory collineation. One might however characterise those Laguerre planes by the existence of a Lie involution of a certain type. A Lie involution of a Laguerre plane at some circle k maps all points not incident with k to circles touching k and vice versa. So it corresponds to an involution of the Lie geometry mapping the point at infinity to the point corresponding to k. If this involution meets the requirements stated in Proposition 4.16, then the Laguerre plane is the sister of a flat Möbius plane.

For every cycle p the map τ induces a natural isomorphism between $\mathfrak{L}(\mathfrak{Mo}, p)$ and $\mathfrak{L}(\mathfrak{Mo}, \tau(p))$, thus we need only specify a circle rather than a cycle for this construction.

We now investigate the properties of the induced involution ϑ in the case where the Laguerre plane is obtained from a flat Möbius plane using a point rather than a cycle. The next lemma links the results obtained in this book with Groh's characterisation of the Laguerre planes obtained from flat Möbius planes ([16]).

(6.3) Lemma: *Suppose ϑ is an involution on a flat Laguerre plane $\mathfrak{L} = (\mathcal{S}, \mathcal{K})$ without fixed points. Then the following conditions are equivalent:*

a) The geometry $\mathfrak{Mo}(\mathfrak{L}, \vartheta)$ is a flat Möbius plane.
b) There are two fixed circles which intersect.
c) There is a circle having no point in common with its image under ϑ.

Proof: Equivalence of the first two conditions is shown in Theorem 6.2. If $\mathfrak{Mo}(\mathfrak{L}, \vartheta)$ is a flat Möbius plane $\mathfrak{Mo}$, then there is a point s of $\mathfrak{Mo}$ such that $\mathfrak{L} = \mathfrak{L}(\mathfrak{Mo}, s)$ and ϑ is the induced involution. Since conditions a) and b) are equivalent, there is a circle k fixed by ϑ. The space $\mathcal{S} \setminus k$ splits into two connected components which get interchanged by ϑ as we have seen in the proof of Proposition 4.16. Let l denote a circle having no point in common with k ([13]: 3.3). Then l lies entirely in one component $\mathcal{S} \setminus k$ and $\vartheta(l)$ in the other, hence $l \cap \vartheta(l) = \emptyset$. This proves the implication a) $\rightarrow$ c).

Finally, suppose there is a circle l such that $l \cap \vartheta(l) = \emptyset$. Let a denote a point on l. Then $\vartheta(a)$ is not parallel to a, for otherwise the parallel class $\Gamma(a) = \Gamma(\vartheta(a)) \cong \mathbb{R}$ would contain a point fixed by ϑ. Let A denote the space of circles through a and $\vartheta(a)$. Then A is homeomorphic to $\mathbb{R}$ and ϑ induces an involution on A. Since every involution on $\mathbb{R}$ has a fixed point, as can be deduced from the intermediate value theorem, there is a circle k through a and $\vartheta(a)$ fixed by ϑ.

The space $\mathcal{S} \setminus (l \cup \vartheta(l))$ consists of three connected components. One of these components is relatively compact, while the other two are not. Let C and D denote the two connected components which are not relatively compact. Then ϑ interchanges C and D while it leaves the relatively compact connected component invariant. The space $\mathcal{S} \setminus k$ consists of two connected components E and F and either $E \cap C$ or $E \cap D$ is not relatively compact. Moreover, if $E \cap C$ is not relatively compact, then $E \cap D$ is relatively compact and therefore $F \cap D$ is not relatively compact. So let us assume that $E \cap C$ is not relatively compact and let $b \in E \cap C$. Then $\vartheta(b)$ has to be in $F \cap D$. In particular, the points b and $\vartheta(b)$ are in different connected components of $\mathcal{S} \setminus k$, therefore every circle through b and $\vartheta(b)$ intersects k. On the other hand, just as above one sees that there is a circle through b and $\vartheta(b)$ fixed by ϑ. Hence, we have two intersecting circles fixed by ϑ. This proves the implication c) $\to$ b). ❀

The construction of a sister of a Möbius plane uses a quadrangle as an intermediate step. As for the sister of a Laguerre plane one can however describe the sister of a Möbius plane also directly.

If s is a point of a flat Möbius plane $\mathfrak{Mo}$, then the points of the Laguerre plane $\mathfrak{L}(\mathfrak{Mo}, s)$ are precisely the cycles incident with s and the circles of the Laguerre plane $\mathfrak{L}(\mathfrak{Mo}, s)$ are all points of the Möbius plane different from s and all cycles not incident with s. Two points of the Laguerre plane are parallel if and only if the corresponding cycles of the Möbius plane touch in s, and a flag of the Laguerre plane consists either of a pair of touching cycles, one incident with s and the other not, or of a cycle through s and a point incident with this cycle. Equivalently, if we work in the derived affine plane $\mathfrak{A}(\mathfrak{Mo}, s)$, then the points of the Laguerre plane are just the oriented lines of the affine plane and circles of the Laguerre plane are the points and oriented circles of the derived affine plane. In the classical case this is the original model of a Laguerre plane.

If k is a cycle of a flat Möbius plane $\mathfrak{Mo}$, then the points of the Laguerre plane $\mathfrak{L}(\mathfrak{Mo}, k)$ are precisely the cycles different from k that touch k plus all points incident with k. Circles of the Laguerre plane $\mathfrak{L}(\mathfrak{Mo}, k)$ are all points of the Möbius plane not incident with k plus all cycles not touching k. A point of the Laguerre plane is incident with a circle, if and only if the corresponding objects in the Möbius plane are parallel or incident or if the corresponding objects in the Möbius plane touch.

6.3 Sisters of Minkowski planes

Now consider a locally compact Minkowski plane $\mathfrak{Mi}$ with parameter t and construct a quadrangle $\mathfrak{GQ} = (\mathcal{P}, \mathcal{L})$ via the lifted Lie geometry of that Minkowski plane. By Theorem 5.26 this quadrangle is antiregular and compact with parameter t. Thus, for every point p of this quadrangle the derived structure $\mathfrak{L}(\mathfrak{G}\mathfrak{Q}, p)$ yields a locally compact Laguerre plane with parameter t. Again, we will call the Laguerre plane $\mathfrak{L}(\mathfrak{Mi}, p) := \mathfrak{L}(\mathfrak{G}\mathfrak{Q}(\mathfrak{Mi}), p)$ the sister of $\mathfrak{Mi}$ at p.

If p is a point of the Minkowski plane rather than a cycle, then the involution τ acting on the quadrangle which is induced by the covering transformation on $\widetilde{\mathcal{K}}$ acts also on the derived Laguerre plane, since τ fixes p. As before, we will denote the involution on the Laguerre plane by ϑ. The properties of τ stated in Theorem 5.30 and Theorem 5.37 imply that the involution ϑ leaves precisely two parallel classes fixed, and that these two parallel classes are even fixed pointwise. Moreover, the space of tangent pencils fixed by ϑ consists of two connected components, each of which is homeomorphic to $\mathbb{R}^t$ and every tangent pencil in one connected component meets every tangent pencil in the other connected component. Also, every fixed point is the support of precisely one tangent pencil fixed by ϑ. This is equivalent to the condition that for every pair of non-parallel fixed points there is precisely one fixed circle containing these two points.

If a locally compact Laguerre plane with parameter 2 satisfies the above conditions, then one can even reconstruct the original Minkowski plane. For a locally compact Laguerre plane with parameter 1 one can also reverse the construction by methods of Corollary 5.33, but the Minkowski plane is not uniquely determined.

If p is a cycle of the Minkowski plane, then the map τ does not induce a collineation on the derived Laguerre plane. Thus, there is no handy way to characterise the Laguerre planes obtained in this way except saying that the quadrangle constructed from the Laguerre plane satisfies the conditions of Theorem 5.30 and Theorem 5.37.

We will summarise the most important of the above results in the following theorems.

For Minkowski planes and Laguerre planes with parameter 2 we have:

(6.4) Theorem: *Suppose p is a point or a cycle of a locally compact Minkowski plane $\mathfrak{Mi}$ with parameter 2. Then*

$$\mathfrak{L}(\mathfrak{Mi}, p) = \mathfrak{L}(\mathfrak{G}\mathfrak{Q}(\mathfrak{Mi}), p)$$

is a locally compact Laguerre plane with parameter 2.

Suppose p is a point of $\mathfrak{Mi}$. Then the involution $\vartheta_{\mathfrak{Mi}}$ induced on $\mathfrak{L}(\mathfrak{Mi}, p)$ by reversing orientation of cycles is an involutory collineation fixing precisely two parallel *classes, which are even fixed pointwise, such that through any pair of non-parallel fixed point there exists precisely one fixed circle.*

Conversely, if $\mathfrak{L}$ is a locally compact Laguerre plane with parameter 2 and with an involution ϑ as above, then

$$\mathfrak{Mi}(\mathfrak{L}, \vartheta) := \mathfrak{Mi}(\mathfrak{G}\mathfrak{Q}(\mathfrak{L}), \tau),$$

where τ is the involution on $\mathfrak{G}\mathfrak{Q}(\mathfrak{L})$ induced by ϑ, is a locally compact Minkowski plane with parameter 2, called the **brother** *of $\mathfrak{Mi}$.*

In addition, the above constructions are inverse to each other, i.e., if ϑ is an involution on a locally compact Laguerre plane with parameter 2 and with the above properties, then

$$\mathfrak{L}(\mathfrak{Mi}(\mathfrak{L}, \vartheta), \infty_{\mathfrak{L}}) = \mathfrak{L} \qquad \text{and} \qquad \vartheta_{\mathfrak{Mi}(\mathfrak{L},\vartheta)} = \vartheta.$$

Conversely,

$$\mathfrak{Mi}(\mathfrak{L}(\mathfrak{Mi}, p), \vartheta_{\mathfrak{Mi}}) = \mathfrak{Mi} \qquad \text{and} \qquad \infty_{\mathfrak{L}(\mathfrak{Mi},p)} = p$$

for every point or cycle p of a locally compact Minkowski plane $\mathfrak{Mi}$ with parameter 2. ⎊

For flat Minkowski planes and flat Laguerre planes we have the following result.

(6.5) Theorem: *If p is a point or a cycle of a flat Minkowski plane $\mathfrak{Mi}$, then*

$$\mathfrak{L}_i(\mathfrak{Mi}, p) := \mathfrak{L}(\mathfrak{G}\mathfrak{Q}_i(\mathfrak{Mi}), p),$$

where $i = 1, 2$, is a flat Laguerre plane, called the sister of $\mathfrak{Mi}$ at p.

Suppose p is a point of $\mathfrak{Mi}$. Then the involution $\vartheta_{\mathfrak{Mi}}$ induced on $\mathfrak{L}_i(\mathfrak{Mi}, p)$ by reversing orientation of cycles is an involutory collineation fixing precisely two parallel classes, which are even fixed pointwise, such that through any pair of non-parallel fixed points there exists precisely one fixed circle.

Conversely, every flat Laguerre plane with an involution as above is the sister of a flat Minkowski plane. ⎊

7 The problem of Apollonius

Introduction

Antiregularity of generalised quadrangles is very closely connected to the so called "problem of Apollonius". Unfortunately, the original two books by Apollonius of Perga concerned with this problem are lost. But due to Pappos some of the material covered in these two books survived. According to [19], the problem as posed and solved by Apollonius of Perga reads as follows:

> *Given three things, each of which may be a point, a straight line, or a circle, to draw a circle which shall pass through the point or points (if such are given) and touch the straight lines or circles, as the case may be.*

In [7] Coxeter gives a nice survey of some aspects of the classical problem of Apollonius. The problem has attracted the attention of emperors and princesses (see [7]). The problem of Apollonius can also be stated in the Lie geometry of the classical Möbius plane. It then reads:

> *Given three points of the Lie geometry of the Möbius plane, construct a circle of the Möbius plane that is collinear with the three given points in the Lie geometry.*

In the non-classical context it makes no sense to ask for the 'construction' of a circle. However, one still can determine the number of circles collinear with three given points of the Lie geometry. Also, the classical problem is restricted to Möbius planes, since circles are natural objects for ruler and compass constructions but parabolas and hyperbolas are not (unless, instead of a compass, one uses a gadget that allows the construction of parabolas or hyperbolas). This restriction is no longer valid if we are only interested in the number of touching circles. So, a more general version of the problem of Apollonius can be stated as follows:

> *Given three points of the Lie geometry of a circle plane, determine the number of circles of the circle plane collinear with the three given points in the Lie geometry.*

For flat Laguerre planes this problem has been solved using the fact that the Lie geometry of a flat Laguerre plane is an antiregular quadrangle ([38]). For locally compact Laguerre planes with parameter 2 this problem has been solved directly ([42]) thereby showing that the quadrangle constructed from a locally compact Laguerre plane with parameter 2 is antiregular. Now we are in a different position. In Chapter 3 it was shown that the quadrangle constructed from a locally compact Laguerre plane with parameter 2 is antiregular. We can therefore use this information to solve the problem of Apollonius in locally compact Laguerre planes with parameter 2 analogously to the flat case.

In both cases the answer is that the number of circles touching three given circles depends only on the topological situation, hence it is always the same as in the corresponding classical case. Not surprisingly, the answer for Möbius planes and Minkowski planes is analogous and the methods used in the proof are almost the same. The main difference is that in the case of Möbius planes or Minkowski planes we have to work with the lifted Lie geometry rather than with the Lie geometry. But this will create little trouble.

There is a nice article by Bruen, Fisher and Wilker ([5]) on the number of common touching circles of three given circles or points in the classical Möbius plane. They mainly work in the derived affine plane at some suitable point s and, implicitly, use also topological arguments. Recently, J.Ch. Fisher has discussed the problem of Apollonius for circle planes over fields ([10]). He also links the solution of the problem to properties of the associated quadrangle. The problem of Apollonius has been partly solved by H. J. Groh for Möbius planes with parameter 1 in [17] and for Laguerre planes with parameter 1 in [14].

The problem of Apollonius, or at least our version of it, lacks symmetry. The given objects can be circles or points of the circle plane but we are only interested in common touching circles and not in common points. This lack of symmetry makes the solution of the problem slightly more complicated but also more interesting.

(7.1) Theorem: *Let $\mathfrak{L} = (\mathcal{S}, \mathcal{K})$ be a finite-dimensional connected Laguerre plane. Then Table 1 gives the number of circles collinear with three given elements of $\mathcal{S} \cup \mathcal{K}$ for all possible configurations. The column headed n states the number of touching circles in the flat case and the column headed p2 indicates whether or not this configuration occurs also in the parameter 2 case. If the configuration occurs also in the parameter 2 case, then the number of touching circles is the same as in the flat case.*

Let $\mathfrak{Mo} = (\mathcal{S}, \mathcal{K})$ be a flat Möbius plane. Then Table 2 gives the number of

	n	$p2$	
two parallel points plus one element of $\mathcal{S} \cup \mathcal{K}$	0	+	•
three pairwise non-parallel points	1	+	•
two non-parallel points s_1, s_2; one circle k			
$s_1 \in k$	1	+	•
s_1, s_2 in different connected components of $\mathcal{S} \setminus k$	0	−	
s_1, s_2 in the same connected component of $\mathcal{S} \setminus k$	2	+	
one point s; two circles k_1, k_2			
$k_1 \cap k_2 = \{s\}$	∞	+	•
$k_1 \cap k_2 = \{u\}$ and $s \nparallel u$	1	+	•
$k_1 \cap k_2 = \{u\} \neq \{s\}$ and $s \parallel u$	0	+	•
$\|k_1 \cap k_2\| \neq 1$; $s \in k_1 \setminus k_2$	1	+	
$\|k_1 \cap k_2\| = 2$; $s \in k_1 \cap k_2$	0	+	•
$\|k_1 \cap k_2\| = 2$; $\|k_1 \cap k_2 \cap \Gamma(s) \setminus \{s\}\| = 1$	1	+	
$\|k_1 \cap k_2\| = 2$; s in one of the relatively compact connected components of $\mathcal{S} \setminus (k_1 \cup k_2)$	0	−	
$\|k_1 \cap k_2\| = 2$; $k_1 \cap k_2 \cap \Gamma(s) = \emptyset$; s not in a relatively compact connected component of $\mathcal{S} \setminus (k_1 \cup k_2)$	2	+	
$k_1 \cap k_2 = \emptyset$; s not in the relatively compact component of $\mathcal{S} \setminus (k_1 \cup k_2)$	0	−	
$k_1 \cap k_2 = \emptyset$; s in the relatively compact connected component of $\mathcal{S} \setminus (k_1 \cup k_2)$	2	−	
three circles k_1, k_2, k_3			
paarwise disjoint	0	−	
$k_1 \cap k_2 = \{u\} = k_1 \cap k_3$	∞	+	
$k_1 \cap k_2 = \{u\} \neq k_1 \cap k_3$; $u \in k_1 \cap k_3$	0	+	
$k_1 \cap k_2 = \{u\} \not\subset k_1 \cap k_3$	1	+	
$\|k_1 \cap k_2\| = 2 = \|k_1 \cap k_3\|$; $k_2 \cap k_3 = \emptyset$	0	−	
$\|k_1 \cap k_2\| = 2$; $\|k_1 \cap k_3\| = \|k_2 \cap k_3\| \neq 1$	2	+	
$\|k_i \cap k_j\| = 2, i \neq j$; $\|k_1 \cap k_2 \cap k_3\| = 1$	1	+	
$\|k_i \cap k_j\| = 2, i \neq j$; $\|k_1 \cap k_2 \cap k_3\| = 2$	0	+	

Table 1 — Laguerre planes

	n	
three points s_1, s_2, s_3	1	•
two points s_1, s_2; one circle k		
$s_1 \in k$	1	•
s_1, s_2 in the same connected component of $\mathcal{S} \setminus k$	2	
s_1, s_2 in different connected components of $\mathcal{S} \setminus k$	0	
one point s; two circles k_1, k_2		
$k_1 \cap k_2 = \{s\}$	∞	•
$s \in k_1 \setminus k_2$	2	
$k_1 \cap k_2 = \{x\} \neq \{s\}$; s in the connected component of $\mathcal{S} \setminus (k_1 \cup k_2)$ bounded by $k_1 \cup k_2$	3	
$k_1 \cap k_2 = \{x\} \neq \{s\}$; s not in the connected component of $\mathcal{S} \setminus (k_1 \cup k_2)$ bounded by $k_1 \cup k_2$	1	
$\lvert k_1 \cap k_2 \rvert = 2$; $s \in k_1 \cap k_2$	0	•
$\lvert k_1 \cap k_2 \rvert = 2$; $s \notin k_1 \cup k_2$	2	
$k_1 \cap k_2 = \emptyset$; $s \notin k_i$; s in the connected component of $\mathcal{S} \setminus (k_1 \cup k_2)$ bounded by $k_1 \cup k_2$	4	
$k_1 \cap k_2 = \emptyset$; $s \notin k_i$; s not in the connected component of $\mathcal{S} \setminus (k_1 \cup k_2)$ bounded by $k_1 \cup k_2$	0	
three circles k_1, k_2, k_3		
pairwise disjoint; k_i separates k_j and k_l for all $\{i, j, l\} = \{1, 2, 3\}$	8	
pairwise disjoint; k_1 separates k_2 and k_3	0	
$\lvert k_1 \cap k_2 \cap k_3 \rvert = 2$	0	
$\lvert k_1 \cap k_2 \cap k_3 \rvert = 1 = \lvert k_i \cap k_j \rvert, i \neq j$	∞	
$\lvert k_1 \cap k_2 \cap k_3 \rvert = 1 = \lvert k_1 \cap k_3 \rvert$; $\lvert k_1 \cap k_2 \rvert = \lvert k_2 \cap k_3 \rvert = 2$	2	
$\lvert k_1 \cap k_2 \cap k_3 \rvert = 1$; $\lvert k_i \cap k_j \rvert = 2$, $i \neq j$	4	
$k_1 \cap k_2 \cap k_3 = \emptyset$; $\lvert k_i \cap k_j \rvert = 2$, $i \neq j$	8	
$k_1 \cap k_2 \cap k_3 = \emptyset$; $\lvert k_1 \cap k_2 \rvert = \lvert k_2 \cap k_3 \rvert = 2$; $\lvert k_1 \cap k_3 \rvert = 1$	6	
$k_1 \cap k_2 \cap k_3 = \emptyset$; $\lvert k_1 \cap k_2 \rvert = 2$; $\lvert k_1 \cap k_3 \rvert = \lvert k_2 \cap k_3 \rvert = 1$	5	
$k_1 \cap k_2 \cap k_3 = \emptyset$; $\lvert k_i \cap k_j \rvert = 1$, $i \neq j$	5	
$k_1 \cap k_3 = \emptyset = k_1 \cap k_3$; $\lvert k_1 \cap k_2 \rvert = 1$; k_2 separates k_1 and k_3	2	
$k_1 \cap k_3 = \emptyset = k_1 \cap k_3$; $\lvert k_1 \cap k_2 \rvert = 1$; k_2 does not separate k_1 and k_3	6	
$k_1 \cap k_3 = \emptyset$; $\lvert k_1 \cap k_2 \rvert = 2$	4	
$k_1 \cap k_3 = \emptyset$; $\lvert k_1 \cap k_2 \rvert = \lvert k_2 \cap k_3 \rvert = 1$; k_2 does not separate k_1 and k_3	5	
$k_1 \cap k_3 = \emptyset$; $\lvert k_1 \cap k_2 \rvert = \lvert k_2 \cap k_3 \rvert = 1$; k_2 separates k_1 and k_3	3	

Table 2 — Möbius planes

	n	$p2$	
two parallel points plus one element of $\mathcal{S} \cup \mathcal{K}$ or two circles contained in different connected components of $\mathcal{K}$ plus one element of $\mathcal{S} \cup \mathcal{K}$	0	+	•
three mutually non-parallel points	1	+	
two non-parallel points s_1, s_2; one circle k			
$s_1 \in k$	1	+	•
$\{\delta(s_1, s_2), \delta(s_2, s_1)\} \cap k = \emptyset$; $\Gamma(s_1) \cup k$ does not separate s_2 and $\delta(\gamma_-(s_1, k), \gamma_+(s_1, k))$.	2	+	
$\{\delta(s_1, s_2), \delta(s_2, s_1)\} \cap k = \emptyset$; $\Gamma(s_1) \cup k$ separates s_2 and $\delta(\gamma_-(s_1, k), \gamma_+(s_1, k))$.	0	−	
$\delta(s_1, s_2) \in k$; $\delta(s_2, s_1) \notin k$	1	+	
$\{\delta(s_1, s_2), \delta(s_2, s_1)\} \subset k$	0	+	
one point s; two circles k_1, k_2 contained in the same connected component of $\mathcal{K}$			
$k_1 \cap k_2 = \{s\}$	∞	+	•
$k_1 \cap k_2 = \{x\}$; $s \notin \Gamma(x)$	1	+	•
$k_1 \cap k_2 = \{x\}$; $s \in \Gamma(x) \setminus \{x\}$	0	+	•
$\lvert k_1 \cap k_2 \rvert = 2$; $s \in k_1 \cap k_2$	0	+	•
$\lvert k_1 \cap k_2 \rvert = 2$; $\lvert k_1 \cap k_2 \cap \Gamma(s) \rvert = 2$	0	+	
$\lvert k_1 \cap k_2 \rvert = 2$; $\lvert k_1 \cap k_2 \cap \Gamma(s) \setminus \{s\} \rvert = 1$	2	+	
$\lvert k_1 \cap k_2 \rvert = 2$; $\lvert k_1 \cap k_2 \cap \Gamma(s) \rvert = 0$; s in the component of $\mathcal{S} \setminus (k_1 \cup k_2)$ containing points parallel to $k_1 \cap k_2$	4	+	
$\lvert k_1 \cap k_2 \rvert = 2$; $\lvert k_1 \cap k_2 \cap \Gamma(s) \rvert = 0$; $s \notin k_i$; s not in the component of $\mathcal{S} \setminus (k_1 \cup k_2)$ containing points parallel to $k_1 \cap k_2$	0	−	
$s \in k_1$; $s \notin k_2$	2	+	
$k_1 \cap k_2 = \emptyset$; $s \notin k_i$	2	−	

Table 3a — Minkowski planes

circles collinear with three given elements of $\mathcal{S} \cup \mathcal{K}$ for all possible configurations. The column headed n states the number of touching circles.

Let $\mathfrak{Mi} = (\mathcal{S}, \mathcal{K})$ be a finite-dimensional connected Minkowski plane. Then Table 3a and Table 3b give the number of circles collinear with three given elements of $\mathcal{S} \cup \mathcal{K}$ for all possible configurations. The column headed n states the number of touching circles in the flat case and the column headed p2 indicates whether or not this configuration occurs also in the parameter 2 case. If the configuration occurs also in the parameter 2 case, then the number of touching circles is the same as in

three circles k_1, k_2, k_3 contained in the same connected component of $\mathcal{K}$	n	$p2$
pairwise disjoint	0	$-$
$\lvert k_1 \cap k_2 \cap k_3 \rvert = 2$	0	$+$
$\lvert k_1 \cap k_2 \cap k_3 \rvert = 1 = \lvert k_i \cap k_j \rvert, i \neq j$	∞	$+$
$\lvert k_1 \cap k_2 \cap k_3 \rvert = 1 = \lvert k_1 \cap k_3 \rvert$; $\lvert k_1 \cap k_3 \rvert = \lvert k_1 \cap k_3 \rvert = 2$	2	$+$
$\lvert k_1 \cap k_2 \cap k_3 \rvert = 1$; $\lvert k_i \cap k_j \rvert = 2$, $i \neq j$	4	$+$
$k_1 \cap k_2 \cap k_3 = \emptyset$; $\lvert k_i \cap k_j \rvert = 2$, $i \neq j$; k_1 meets k_3 in the same connected component of $k_3 \setminus \Gamma(s)$ as k_2, where $s \in k_1 \cap k_2$	8	$+$
$k_1 \cap k_2 \cap k_3 = \emptyset$; $\lvert k_i \cap k_j \rvert = 2$, $i \neq j$; k_1 meets k_3 not in the same connected component of $k_3 \setminus \Gamma(s)$ as k_2, where $s \in k_1 \cap k_2$	0	$-$
$k_1 \cap k_2 \cap k_3 = \emptyset$; $\lvert k_1 \cap k_2 \rvert = \lvert k_2 \cap k_3 \rvert = 2$; $\lvert k_1 \cap k_3 \rvert = 1$; k_1 meets k_3 in the same connected component of $k_3 \setminus \Gamma(s)$ as k_2, where $s \in k_1 \cap k_2$	6	$+$
$k_1 \cap k_2 \cap k_3 = \emptyset$; $\lvert k_1 \cap k_2 \rvert = \lvert k_2 \cap k_3 \rvert = 2$; $\lvert k_1 \cap k_3 \rvert = 1$; k_1 meets k_3 not in the same connected component of $k_3 \setminus \Gamma(s)$ as k_2, where $s \in k_1 \cap k_2$	2	$-$
$k_1 \cap k_2 \cap k_3 = \emptyset$; $\lvert k_1 \cap k_2 \rvert = 2$; $\lvert k_1 \cap k_3 \rvert = \lvert k_2 \cap k_3 \rvert = 1$; k_1 meets k_3 in the same connected component of $k_3 \setminus \Gamma(s)$ as k_2, where $s \in k_1 \cap k_2$	5	$+$
$k_1 \cap k_2 \cap k_3 = \emptyset$; $\lvert k_1 \cap k_2 \rvert = 2$; $\lvert k_1 \cap k_3 \rvert = \lvert k_2 \cap k_3 \rvert = 1$; k_1 meets k_3 not in the same connected component of $k_3 \setminus \Gamma(s)$ as k_2, where $s \in k_1 \cap k_2$	3	$-$
$k_1 \cap k_2 \cap k_3 = \emptyset$; $\lvert k_i \cap k_j \rvert = 1$, $i \neq j$;	3	5
$k_1 \cap k_2 = \emptyset$; $\lvert k_1 \cap k_2 \rvert = 2$,	4	$-$
$k_1 \cap k_2 = \emptyset$; $\lvert k_1 \cap k_2 \rvert = \lvert k_2 \cap k_3 \rvert = 1$,	3	$-$

Table 3b — Minkowski planes cont.

the flat case except for one configuration, where consequently the correct number is indicated.

Proof: The cases marked • follow directly from the axioms or are trivial and may be checked by the reader. Most of the remaining cases are treated best in the quadrangle constructed from the circle plane. Apart from antiregularity of these quadrangles square projections will play a prominent rôle, especially in the parameter 1 case. The main problem there is to find a suitable square projection $\mu_{p_1,p_2,g}$, i.e., suitable points p_1 and p_2 as well as a suitable line g for the configuration considered. We have seen in Chapter 2 that in the parameter 1 case the truncated line $g \setminus \{\pi(p_1, g), \pi(p_2, g)\}$ consists of two connected components. One of these connected components is not in the image of the square projection while every point in the other connected component has precisely two preimages under the square projection. We will use this result frequently without reference. In most cases the problem will be to determine which of the two components is the one in the image of the square projection.

The arguments for the different configurations run very much along the same lines—with subtle variations. To grasp the main idea the reader needs only to look at some of the cases, preferably at some of the configurations in flat Möbius planes. A full understanding can, unfortunately, only be gained by contemplating the whole and rather lengthy proof.

The proof is split into three parts, one for each type of circle planes.

7.1 The problem of Apollonius in Laguerre planes

The solution to the problem of Apollonius in finite-dimensional connected Laguerre planes has already been published in [38] for the parameter 1 case and in [42] for the parameter 2 case. The proof for the parameter 1 case already uses antiregularity of the quadrangle constructed from the Laguerre plane and, implicitly, properties of square projections. However, for sake of completeness, we will reproduce it here. The proof for the parameter 2 case presented here uses new ideas.

To be able to distinguish between a point or a circle of the Laguerre plane $\mathfrak{L}$ and the corresponding point of the generalised quadrangle $\mathfrak{GQ}(\mathfrak{L})$, we introduce the identification $\iota : \mathcal{P} \to \mathcal{S} \cup \mathcal{K} \cup \{\infty_\mathfrak{L}\} : x \mapsto x$ where $\mathcal{P} = \mathcal{S} \cup \mathcal{K} \cup \{\infty_\mathfrak{L}\}$ is the point space of the quadrangle $\mathfrak{GQ}(\mathfrak{L})$.

Suppose s_1 and s_2 are two non-parallel points and the circle k is neither incident with s_1 nor incident with s_2. We first consider the parameter 1 case. Let $g := \infty_\mathfrak{L} \vee \iota^{-1}(s_2)$ and consider the square projection $\mu := \mu_{\iota^{-1}(s_1), \iota^{-1}(k), g}$ from $\iota^{-1}(s_1)^\perp \cap \iota^{-1}(k)^\perp$ to g. One of the two connected components of the truncated line $g \setminus \{\infty_\mathfrak{L} = \pi(\iota^{-1}(s_1), g), t := \pi(\iota^{-1}(k), g)\}$ is not in the image of μ while every point of the other connected component has precisely two preimages under μ. In the Laguerre plane the truncated line $g \setminus \{\infty_\mathfrak{L}, t\}$ corresponds to $\Gamma(s_2) \setminus \{\iota(t)\} = \Gamma(s_2) \setminus k$. The two connected components of $\Gamma(s_2) \setminus k$ are contained in different connected components of $\mathcal{S} \setminus k$. Now, if s is a point on $\Gamma(s_2)$ such that k separates

s and s_1, then, by connectedness arguments, there is no circle through s_1 touching k and incident with s. But then for every point s of the connected component of $\Gamma(s_2) \setminus k$ contained in the same connected component of $\mathcal{S} \setminus k$ as s_1 the preimage $\mu^{-1}(\iota^{-1}(s)) = \iota^{-1}(s_1)^\perp \cap \iota^{-1}(s)^\perp \cap \iota^{-1}(k)^\perp$ consists of two points x and y. If the Laguerre plane has parameter 2, then the quadrangle $\mathfrak{GQ}(\mathfrak{L})$ is centric antiregular. Thus for every point $s \in \Gamma(s_2) \setminus k$ there are also precisely two points x and y in the centre $\iota^{-1}(s_1)^\perp \cap \iota^{-1}(s)^\perp \cap \iota^{-1}(k)^\perp$. Since s_1 is not parallel to s_2 we obtain that $\iota(x)$ and $\iota(y)$ are circles. Thus, for both parameters, $\iota(x)$ and $\iota(y)$ are the two circles through s and s_1 touching k.

The proof for the next two cases works for both parameters. If k_1 and k_2 are two non-tangent circles and s is a point incident with k_1 but not incident with k_2, then $\iota(\pi(\iota^{-1}(k_2), \iota^{-1}(s) \vee \iota^{-1}(k_1)))$ is the unique circle through s tangent to k_1 and k_2.

Suppose k_1 and k_2 are two properly intersecting circles and s is a point not contained in $k_1 \cup k_2$ but parallel to a point t in $k_1 \cap k_2$. Then $\{\iota^{-1}(s), \iota^{-1}(k_1), \iota^{-1}(k_2)\}$ forms a triad and $\iota^{-1}(t) \in \iota^{-1}(s)^\perp \cap \iota^{-1}(k_1)^\perp \cap \iota^{-1}(k_2)^\perp$. Thus, by antiregularity, there is precisely one other point x in the centre $\iota^{-1}(s)^\perp \cap \iota^{-1}(k_1)^\perp \cap \iota^{-1}(k_2)^\perp$. The circle $\iota(x)$ is the unique circle through s touching k_1 and k_2.

Suppose k_1 and k_2 are two non-tangent circles and s is a point neither incident with k_1 nor k_2 nor parallel to any point in $k_1 \cap k_2$. We first consider the parameter 1 case. Let $g := \infty_{\mathfrak{L}} \vee \iota^{-1}(s)$ and consider the square projection $\mu := \mu_{\iota^{-1}(k_1), \iota^{-1}(k_2), g}$ from $\iota^{-1}(k_1)^\perp \cap \iota^{-1}(k_2)^\perp$ to g. If k_1 and k_2 intersect, then $\infty_{\mathfrak{L}}$ has two preimages under μ, namely the points corresponding to the intersection of k_1 and k_2. If k_1 and k_2 do not intersect, then $\infty_{\mathfrak{L}}$ is not in the image of μ. Thus, if k_1 and k_2 intersect, then every point in the connected component of $g \setminus \{\pi(\iota^{-1}(k_1), g), \pi(\iota^{-1}(k_2), g)\}$ containing $\infty_{\mathfrak{L}}$ has two preimages under μ while the points in the other connected components are not in the image of μ. If k_1 and k_2 do not intersect, then it is just the other way round. In the Laguerre plane, the connected component of $g \setminus \{\pi(\iota^{-1}(k_1), g), \pi(\iota^{-1}(k_2), g)\}$ not containing $\infty_{\mathfrak{L}}$ corresponds to the relatively compact connected component of $\Gamma(s) \setminus \{k_1 \cup k_2\}$. Moreover, the number of circles through s touching k_1 and k_2 corresponds to the number of elements in the preimage of $\iota^{-1}(s)$ under μ. In the parameter 2 case the centre of the triad $\{\iota^{-1}(k_1), \iota^{-1}(k_2), \iota^{-1}(s)\}$ contains precisely two points x and y. Since s is not parallel to any of the two points in $k_1 \cap k_2$, we know that $\iota(x)$ and $\iota(y)$ are circles. Thus $\iota(x)$ and $\iota(y)$ are the two circles through s tangent to k_1 and k_2.

We now come to the cases where three circles are given. If the three circles are pairwise disjoint, then, by connectedness arguments, there is no common tangent circle.

If k_1, k_2 are two tangent circles, then $g := \iota^{-1}(k_1) \vee \iota^{-1}(k_2)$ is a line of the quadrangle. If k_3 is a circle that touches k_1 in the same point as k_2, then $\iota^{-1}(k_3) \in g$ and all circles of the touching pencil spanned by k_1 and k_2 are also tangent to k_3.

If k_3 is a circle that does not touch k_1 in the same point as k_2, then $\iota^{-1}(k_3) \notin g$. Let $x := \pi(\iota^{-1}(k_3), g)$. If $k_1 \cap k_2 \subset k_3$, then $\iota(x)$ is a point, thus there is no circle touching k_1, k_2 and k_3. If $k_1 \cap k_2 \not\subset k_3$, then $\iota(x)$ is the unique circle touching k_1, k_2 and k_3. Incidentally, this configuration was solved directly in Lemma 3.3.

Next, suppose $|k_1 \cap k_2| = 2$ but $k_1 \cap k_2 \cap k_3 = \emptyset$. We first consider the parameter 1 case. Let u denote a point of $k_1 \cap k_2$. In the sister $\mathfrak{L}(\mathfrak{L}, u)$ of $\mathfrak{L}$ at u the circles k_1 and k_2 become points p_1 and p_2 respectively while the circle k_3 becomes a circle q. By the above the situation is as follows: If q separates p_1 and p_2, then there is no circle through p_1 and p_2 touching q, while if q does not separate p_1 and p_2, then there are precisely two circles x and y through p_1 and p_2 touching q. In the original Laguerre plane the circles x and y correspond to circles touching k_1, k_2 and k_3. The problem now is to find out when q separates p_1 and p_2. The circle q separates the point set of $\mathfrak{L}(\mathfrak{L}, u)$ into two connected components. One component corresponds to the set of circles of the original Laguerre plane not intersecting k_3 while the other corresponds to the set of circles of the original Laguerre plane intersecting k_3 properly. So q separates p_1 and p_2 if and only if one of the circles k_1 and k_2 intersects k_3 properly while the other misses k_3. In this case there is no circle touching all three circles k_1, k_2 and k_3. If k_1 and k_2 either both miss k_3 or both intersect k_3 properly, then there are precisely two circles touching k_1, k_2 and k_3. In the parameter 2 case the set $\{\iota^{-1}(k_1), \iota^{-1}(k_2), \iota^{-1}(k_3)\}$ is a triad, thus there are precisely two points x and y in the centre $\iota^{-1}(k_1)^\perp \cap \iota^{-1}(k_2)^\perp \cap \iota^{-1}(k_3)^\perp$. Since $k_1 \cap k_2 \cap k_3 = \emptyset$ these two points correspond to circles in the Laguerre plane, namely to the two circles touching k_1, k_2 and k_3.

Finally, suppose k_1, k_2 and k_3 are three mutually non-tangent circles such that $k_1 \cap k_2 \cap k_3 \neq \emptyset$. Then $\emptyset \neq \iota^{-1}(\{k_1 \cap k_2 \cap k_3\}) \subset \iota^{-1}(k_1)^\perp \cap \iota^{-1}(k_2)^\perp \cap \iota^{-1}(k_3)^\perp$. Thus, by antiregularity, the centre $\iota^{-1}(k_1)^\perp \cap \iota^{-1}(k_2)^\perp \cap \iota^{-1}(k_3)^\perp$ contains two points. Therefore, the number of circles touching k_1, k_2 and k_3 equals $2 - |k_1 \cap k_2 \cap k_3|$.

In a locally compact Laguerre plane with parameter 2 any two circles intersect and the complement of a circle in the point space is still connected. This accounts for the configurations which appear only in the parameter 1 case.

This proves correctness of Table 1.

7.2 The problem of Apollonius in Möbius planes

We now consider the problem of Apollonius in a flat Möbius plane $\mathfrak{Mo}$. Again, we mainly work in the quadrangle $\mathfrak{GQ}(\mathfrak{Mo})$ associated to $\mathfrak{Mo}$. Compared with the situation encountered in the previous section there is one main difference. Instead of an identification of the point space of $\mathfrak{GQ}(\mathfrak{Mi})$ with the point space and space of circles of the circle plane we now have a twofold branched covering $c : \mathcal{P} \to \mathcal{S} \cup \mathcal{K}$. For this reason every configuration in the Möbius plane corresponds to several configurations in the quadrangle.

Our strategy will be as follows: Given three objects $x_1, x_2, x_3 \in \mathcal{S} \cup \mathcal{K}$ of a flat Möbius plane $\mathfrak{Mi}$, we determine the number of points in $p_1^\perp \cap p_2^\perp \cap p_3^\perp$ for the different choices of $p_i \in c^{-1}(x_i)$. Then we determine the number of circles corresponding to these points. If p_1, p_2 and p_3 form a triad, we choose a suitable line g through p_3 and determine whether or not p_3 is in the image of the square projection from $p_1^\perp \cap p_2^\perp$ to g. If p_3 is in the image, then $p_1^\perp \cap p_2^\perp \cap p_3^\perp$ contains precisely two points, namely $\mu^{-1}(p_3)$. If p_3 is not in the image, then $p_1^\perp \cap p_2^\perp \cap p_3^\perp$ is empty.

Let $\tau = \tau_{\mathfrak{Mo}}$ denote the involution on $\mathfrak{GQ}(\mathfrak{Mo})$ induced by reversing orientation of cycles. Since $\tau(p_1^\perp \cap p_2^\perp \cap p_3^\perp) = \tau(p_1)^\perp \cap \tau(p_2)^\perp \cap \tau(p_3)^\perp$ and $c(z) = c(\tau(z))$, we get every circle twice in the above construction. To avoid this, we will omit all cases involving $\tau(p_1)$, where we assume that x_1 is a circle.

The same strategy works basically also for flat Minkowski planes. However, due to the existence of parallel classes considerably more configurations have to be considered.

For the rest of this section p_i denotes an element of $c^{-1}(k_i)$, where k_i is a circle. Thus $c^{-1}(k_i) = \{p_i, \tau(p_i)\}$.

The following result will be very useful in the proof. It says that if we blow up a point to a circle, then the number of common touching circles doubles. More precisely:

(7.2) Lemma: *Suppose k is a circle of a flat Möbius plane $\mathfrak{Mo} = (\mathcal{S}, \mathcal{K})$ and suppose $x_1, x_2 \in \mathcal{S} \cup \mathcal{K}$ such that x_1 and x_2 are contained in a common connected component of $\mathcal{S} \setminus k$. Then for every $s \in k$ the number of circles touching k that touch or are incident with x_1 and x_2 is twice the number of circles incident with s that touch or are incident with x_1 and x_2.*

Proof: Let $p \in c^{-1}(k)$, $q_1 \in c^{-1}(x_1)$ and $q_2 \in c^{-1}(x_2)$ be corresponding objects in the quadrangle $\mathfrak{GQ}(\mathfrak{Mo})$. Let $g := s \vee p$. We have

$$(p^\perp \cap q_1^\perp \cap q_2^\perp) \cap (\tau(p)^\perp \cap q_1^\perp \cap q_2^\perp) = p^\perp \cap \tau(p)^\perp \cap q_1^\perp \cap q_2^\perp \subset k \cap x_1 \cap x_2 = \emptyset.$$

Since $k \cap x_1 \cap x_2 = \emptyset$ the set $c(p^\perp \cap q_1^\perp \cap q_2^\perp)$ contains no points. But then the intersection $(p^\perp \cap q_1^\perp \cap q_2^\perp) \cap \tau\left(\tau(p)^\perp \cap q_1^\perp \cap q_2^\perp\right)$ is empty unless x_1 and x_2 are points rather than circles. However, in this case there is precisely one circle through s, x_1 and x_2 and s is in the image of the square projection of from $q_1^\perp \cap q_2^\perp = x_1^\perp \cap x_2^\perp$ to g. Since k does not separate x_1 and x_2 there is a path in the extended tangent pencil corresponding to g from k to s such that no circle of this path is incident with x_1 or x_2. But then p and s are in a common connected component of $g \setminus \{\pi(q_1, g), \pi(q_2, g)\}$. Thus, also the preimage of p under the square projection contains two points, hence $p^\perp \cap q_1^\perp \cap q_2^\perp = \tau\left(\tau(p)^\perp \cap q_1^\perp \cap q_2^\perp\right)$ contains two points. Consequently, there are precisely two circles through x_1 and x_2 tangent to k.

For the remaining cases we have $(p^\perp \cap q_1^\perp \cap q_2^\perp) \cap \tau\left(\tau(p)^\perp \cap q_1^\perp \cap q_2^\perp\right) = \emptyset$. Hence, we only have to show $\left|p^\perp \cap q_1^\perp \cap q_2^\perp\right| = \left|s^\perp \cap q_1^\perp \cap q_2^\perp\right|$, since then any object in $s^\perp \cap q_1^\perp \cap q_2^\perp$ corresponds to one object in $p^\perp \cap q_1^\perp \cap q_2^\perp$ and one object in $\tau(p)^\perp \cap q_1^\perp \cap q_2^\perp$.

If q_1 is collinear with q_2, then $s^\perp \cap q_1^\perp \cap q_2^\perp = \{\pi(s, q_1 \vee q_2)\}$ and $p^\perp \cap q_1^\perp \cap q_2^\perp = \{\pi(p, q_1 \vee q_2)\}$. Hence $\left|s^\perp \cap q_1^\perp \cap q_2^\perp\right| = \left|p^\perp \cap q_1^\perp \cap q_2^\perp\right| = 1$. If q_1 is not collinear with q_2, we consider the square projection $\mu : q_1^\perp \cap q_2^\perp \to g$. Let $G := c(g)$ denote the extended touching pencil spanned by s and k. Let B denote the set of those elements l of G, such that l has no point in common with the connected component of $\mathcal{S} \setminus k$ containing x_1. Then B is connected and $\{s, k\} \in B$. In fact, B is an interval with endpoints k and s. Furthermore, no element of B touches or is incident with x_1 or x_2. Since the endpoints of $A := c(\mu(q_1^\perp \cap q_2^\perp))$ touch or are incident with x_1 or x_2, we have either $B \subset A$ or $B \cap A = \emptyset$. Thus, the circle k is in A if and only if s is in A and hence $p \in \mu(q_1^\perp \cap q_2^\perp) \iff s \in \mu(q_1^\perp \cap q_2^\perp)$. This implies $\left|p^\perp \cap q_1^\perp \cap q_2^\perp\right| = \left|\mu^{-1}(p)\right| = \left|\mu^{-1}(s)\right| = \left|p^\perp \cap q_1^\perp \cap q_2^\perp\right|$. ♋

Now, continuing the proof of Theorem 7.1, consider the non-trivial cases where two points s_1, s_2 and one circle k are given. If s_1 and s_2 are in the same connected component of $\mathcal{S} \setminus k$, we have seen in the proof of Lemma 7.2 that there are precisely two circles through s_1 and s_2 touching k. If s_1 and s_2 are in different connected components, then, by connectedness arguments, there is no circle through s_1 and s_2 touching k.

One point and two circles

We now consider the case where one point s and two circles k_1 and k_2 are given. If $s \in k_1 \setminus k_2$, let $g := s \vee p_1$. By assumption neither p_2 nor $\tau(p_2)$ is incident with g. Let $v := \pi(p_2, g)$ and $w := \pi(\tau(p_2), g)$. The points v and w are different, since otherwise $v \in g \cap p_2^\perp \cap \tau(p_2)^\perp \in g \cap \mathcal{S} = \{s\}$ which contradicts $s \notin k_2 = p_2^\perp \cap \tau(p_2)^\perp$. Thus $c(v)$ and $c(w)$ are the only two circles through s touching k_1 and k_2. Incidentally, this situation was also treated in Lemma 4.1.

Next suppose $s \notin k_1 \cup k_2$ and $k_1 \cap k_2 = \{x\}$. The space $\mathcal{S} \setminus (k_1 \cup k_2)$ splits into three connected components, one of which is bounded by $k_1 \cup k_2$ while each of the other connected components is already bounded by one of the circles k_1 or k_2. Since k_1 and k_2 touch, we may assume that p_1 and p_2 are collinear, hence $s^\perp \cap p_1^\perp \cap p_2^\perp = \{\pi(s, p_1 \vee p_2)\}$. However, if p_1 and p_2 are collinear, then p_1 and $\tau(p_2)$ are not collinear. Let $g := x \vee \tau(p_2)$ and consider the square projection $\mu : s^\perp \cap p_1^\perp \to g$. The image $A := \mu(s^\perp \cap p_1^\perp)$ is an interval with endpoints $\pi(s, g) = \widetilde{\beta}(x, \tau(p_2), s)$ and $\pi(p_1, g) = x$. The circle $\beta(x, k_2, s) = c(\pi(s, g))$ separates the touching pencil $G := c(g) \setminus \{x\}$ into two connected components X and Y where X consists of all circles $l \neq k_1 \in G$ such that s is in the connected component of $\mathcal{S} \setminus (k_1 \cup l)$ bounded by $k \cup l$, while Y consists of all the circles l in G, such that s is in the connected component of $\mathcal{S} \setminus (k_1 \cup l)$ bounded by already one of the circles k or l. By construction $\tau(p_1) \in g$ and $\mu^{-1}(\tau(p_1)) = s^\perp \cap p_1^\perp \cap \tau(p_1)^\perp = \{s\} \cap k_1 = \emptyset$. Trivially we have $c(\tau(p_1)) = k_1 \in Y$. Thus $c(A) \cap Y = \emptyset$ and $X \subset c(A)$. Hence, if k_2 is a circle touching k_1 such that s is not in the connected component of $\mathcal{S} \setminus (k_1 \cup k_2)$ bounded by $k_1 \cup k_2$, then $\beta(k_1 \cap k_2, k_1, s)$ is the only circle through s touching k_1 and k_2. Conversely, if k_2 is a circle touching k_1 such that s is in the connected component of $\mathcal{S} \setminus (k_1 \cup k_2)$ bounded by $k_1 \cup k_2$, then there are two points u and v in $\mathfrak{GQ}(\mathfrak{Mo})$ such that $s^\perp \cap p_1^\perp \cap p_2^\perp = \{\pi(s, p_1 \vee p_2)\}$ and $s^\perp \cap p_1^\perp \cap \tau(p_2)^\perp = \{u, v\}$. But then $c(\pi(s, p_1 \vee p_2)) = \beta(k_1 \cap k_2, k_1, s)$, $c(u)$ and $c(v)$ are the three circles through s touching k_1 and k_2.

These circles are mutually different as the following arguments show. Suppose $c(u) = c(v)$, then $\mu(u) = \mu(v) \in u^\perp \cap v^\perp \subset \mathcal{S}$, hence $\mu(u) = \mu(v) = g \cap \mathcal{S} = \{x\}$. This implies $u = v = \widetilde{\beta}(x, \tau(p_2), s)$ contradicting $u \neq v$. Also, if $c(v) = \beta(k_1 \cap k_2, k_1, s)$ then, again, $\mu(u) = \mu(v) = x$ which leads to the same contradiction.

For the rest of the proof we will leave it to the reader to check that the circles in question are mutually different. The arguments needed run very much along the above lines.

If $|k_1 \cap k_2| = 2$ and $s \notin k_1 \cup k_2$, let $x \in k_2 \setminus k_1$ such that x and s lie in different connected components of $\mathcal{S} \setminus k_1$. Consider the square projections $\mu : s^\perp \cap p_1^\perp \to g := x \vee p_2$ and $\nu : s^\perp \cap p_1^\perp \to \tau(g) = x \vee \tau(p_2)$. The image $A := \mu(s^\perp \cap p_1^\perp)$ has endpoints $\pi(s,g) = \widetilde{\beta}(x, p_2, s)$ and $\pi(p_1, g)$ while the image $B := \nu(s^\perp \cap p_1^\perp)$ has endpoints $\pi(s, \tau(g)) = \widetilde{\beta}(x, \tau(p_2), s)$ and $\pi(p_1, \tau(g))$. Let $a := c(\pi(s,g)) = c(\pi(s, \tau(g)) = \beta(x, k_2, s)$, $b := c(\pi(p_1, g))$ and $c := c(\pi(p_1, \tau(g)))$ denote the corresponding circles in the extended touching pencil $G := c(g)$ spanned by x and k_2.

Just for a moment consider the derived projective plane $\mathfrak{P}(\mathfrak{Mo}, x)$. There, the extended touching pencil G corresponds to a line pencil $\overline{G}$ through some point on the line at infinity. The lines $\overline{b}$ and $\overline{c}$ corresponding to b and c separate $\overline{G}$ into two connected components. Results of [4], namely that the dual of a topological oval is again a topological oval, imply that one of these components consists of all lines of the line pencil $\overline{G}$ missing $\overline{k}_1$ while the other component consists of all lines of the pencil intersecting $\overline{k}_1$ in two points.

Returning to the Möbius plane this implies that the two circles b and c separate G into two connected components, one of which contains all circles in G intersecting k_1 in two points while the other contains the point x and all circles in G not meeting k_1. Hence b and c lie in different connected components of $G \setminus \{a, x\}$. Since s and x lie in different connected components of $\mathcal{S} \setminus k_1$, we know, by a case already considered, that x is neither in A nor in B. Thus $G \setminus (c(A) \cup c(B))$ is precisely the connected component of $G \setminus \{b, c\}$ containing x and therefore also all the circles missing k_1 while $c(A) \cap c(B) = \{a\}$. Hence k_2 is either in $c(A)$ or $c(B)$ but not in both. We may assume $k_2 \in c(A)$ and therefore $p_2 \in A$ but $\tau(p_2) \notin B$, hence there are points u and v such that $s^\perp \cap p_1^\perp \cap p_2^\perp = \mu^{-1}(p_2) = \{u, v\}$ and $s^\perp \cap p_1^\perp \cap \tau(p_2)^\perp = \mu^{-1}(\tau(p_2)) = \emptyset$. Thus $c(u)$ and $c(v)$ are the two circles through s touching k_1 and k_2.

If $k_1 \cap k_2 = \emptyset$, then again $\mathcal{S} \setminus (k_1 \cup k_2)$ consists of three connected components, one of which is bounded by $k_1 \cup k_2$ while each of the other connected components is already bounded by one of the circles k_1 or k_2. Suppose s is in a connected component bounded by already one circles, k_1 say, then k_2 and s lie in different connected components of $\mathcal{S} \setminus k_1$, hence there is no circle through s touching k_1 and k_2.

Now suppose $k_1 \cap k_2 = \emptyset$ and s is contained in the connected component of $\mathcal{S} \setminus (k_1 \cup k_2)$ bounded by $k_2 \cup k_2$. Then an application of Lemma 7.2 shows that there are precisely four circles through s touching k_1 and k_2.

Three circles

Finally, suppose we are given three circles k_1, k_2 and k_3. If the circles are pairwise disjoint, then Lemma 7.2 together with cases already considered yield the desired results.

If there are two points $y, z \in \mathcal{S}$ such that $k_1 \cap k_2 \cap k_3 = \{y, z\}$, then $p_1^\perp \cap p_2^\perp \cap p_3^\perp = \{y, z\} \subset \mathcal{S}$ for any choice $p_i \in c^{-1}(k_i)$. Thus there is no circle touching touching k_1, k_2 and k_3.

Next suppose there is a point z such that $k_1 \cap k_2 \cap k_3 = \{z\}$. If $k_1 \cap k_2 = k_1 \cap k_3 = k_2 \cap k_3 = \{z\}$, then there are infinitely many circles touching the three given circles, namely all the circles contained in the touching pencil spanned by these three circles. Connectedness arguments show that there are no other circles touching k_1, k_2 and k_3. The case $|k_1 \cap k_2| = 2$ and $|k_1 \cap k_3| = |k_2 \cap k_3| = 1$ cannot occur, since in this situation we may assume that p_3 is collinear with p_1 and p_2. But z is collinear with p_1, p_2 and p_3 which is only possible if z and p_3 are on a line spanned by p_1 and p_2, i.e., if p_1 is collinear with p_2. This implies $k_1 \cap k_2 = \{z\}$, contradicting our assumptions. If $|k_1 \cap k_2| = |k_2 \cap k_3| = 2$ and $|k_1 \cap k_3| = 1$ we may assume that p_1 is collinear with p_3. This implies that for $q \in \{p_2, \tau(p_2)\}$ we have $p_1^\perp \cap q^\perp \cap p_3^\perp = \{\pi(q, p_1 \vee p_3)\} = \{z\}$. However, if p_1 is collinear with p_3, then p_1 is not collinear with $\tau(p_3)$, thus by antiregularity there are cycles x and y, such that $p_1^\perp \cap p_2^\perp \cap \tau(p_3)^\perp = \{x, z\}$ and $p_1^\perp \cap \tau(p_2)^\perp \cap \tau(p_3)^\perp = \{y, z\}$. Then $c(x)$ and $c(y)$ are the two circles touching k_1, k_2 and k_3. If, still under the assumption that there is a point z such $k_1 \cap k_2 \cap k_3 = \{z\}$, we have $|k_i \cap k_j| = 2$ for $i \neq j$, then $z \in p_1^\perp \cap p_2^\perp \cap p_3^\perp \cap \tau(p_2)^\perp \cap \tau(p_3)^\perp$. By antiregularity there are cycles v, w, x and y such that:

$$\begin{aligned} p_1^\perp \cap p_2^\perp \cap p_3^\perp &= \{v, z\}, \\ p_1^\perp \cap p_2^\perp \cap \tau(p_3)^\perp &= \{w, z\}, \\ p_1^\perp \cap \tau(p_2)^\perp \cap p_3^\perp &= \{x, z\}, \\ p_1^\perp \cap \tau(p_2)^\perp \cap \tau(p_3)^\perp &= \{y, z\}. \end{aligned}$$

Thus $c(v)$, $c(w)$, $c(x)$ and $c(y)$ are the four circles touching k_1, k_2 and k_3.

We are now left with the cases where k_1 and k_2 have at least one point in common but there is no point incident with all three circles. We first have a closer look at the square projection in the situation where there is a point $s \in k_1 \cap k_2 \setminus k_3$ and p_1, p_2 and p_3 are pairwise non-collinear. Let $g := s \vee p_1$ denote the line connecting s to p_1 and consider the square projection $\mu : p_2^\perp \cap p_3^\perp \to g : x \mapsto \pi(x, g)$. Again,

$\mu(p_2^\perp \cap p_3^\perp)$ is a closed interval in g with endpoints $\pi(p_2, g) = s$ and $\pi(p_3, g)$. Let $B := \{\pi(t, g) \mid t \in k_3\}$. Then B is an interval with endpoints $\pi(p_3, g)$ and $\pi(\tau(p_3), g)$. Obviously $s \notin B$. Thus either B is contained entirely in $\mu(p_2^\perp \cap p_3^\perp)$ or B is contained entirely in the closure of the complement of $\mu(p_2^\perp \cap p_3^\perp)$ in g. Moreover, if, in addition, there is a point $t \in k_2 \cap k_3$, then $\pi(t, g) = \mu(t) \in B \cap \mu(p_2^\perp \cap p_3^\perp)$. So, unless $\mu(t) = \pi(p_3, g)$ we have $B \subset \mu(p_2^\perp \cap p_3^\perp)$.

Now suppose the intersection $k_1 \cap k_3$ is not empty and that $k_2 \cap k_3$ contains two points. Choose $t \in k_2 \cap k_3$ such that $\mu(t) \neq \pi(p_3, g)$ and let $r \in k_1 \cap k_3$. Then, by the above, $B \subset \mu(p_2^\perp \cap p_3^\perp)$. Moreover $p_1 = \pi(r, g) \in B$, hence $p_1 \in \mu(p_2^\perp \cap p_3^\perp)$ and therefore $p_1^\perp \cap p_2^\perp \cap p_3^\perp = \mu^{-1}(p_1)$ contains precisely two points. This result allows us to treat all the cases still left where any two of the circles k_1, k_2 and k_3 have at least one point in common and one intersection contains two points.

If $|k_1 \cap k_2| = |k_1 \cap k_3| = |k_2 \cap k_3| = 2$, then in the above argument we can replace any of the points p_i by $\tau(p_i)$. Thus there are eight points x_1, x_2, x_3, x_4, y_1, y_2, y_3 and y_4 such that:

$$\begin{aligned} p_1^\perp \cap p_2^\perp \cap p_3^\perp &= \{x_1, y_1\}, \\ p_1^\perp \cap p_2^\perp \cap \tau(p_3)^\perp &= \{x_2, y_2\}, \\ p_1^\perp \cap \tau(p_2)^\perp \cap p_3^\perp &= \{x_3, y_3\}, \\ p_1^\perp \cap \tau(p_2)^\perp \cap \tau(p_3)^\perp &= \{x_4, y_4\}, \end{aligned}$$

and $c(x_1)$, $c(x_2)$, $c(x_3)$, $c(x_4)$, $c(y_1)$, $c(y_2)$, $c(y_3)$ and $c(y_4)$ are the eight circles touching k_1, k_2 and k_3.

If $|k_1 \cap k_2| = |k_2 \cap k_3| = 2$ but $|k_1 \cap k_3| = 1$ we may assume that p_1 is collinear with $\tau(p_3)$, hence not with p_3. This implies that in the above argument we can only replace p_2 by $\tau(p_2)$ or p_1 and p_3 by $\tau(p_1)$ and $\tau(p_3)$ simultaneously. Thus there are four points x_1, x_2, y_1 and y_2 such that:

$$\begin{aligned} p_1^\perp \cap p_2^\perp \cap p_3^\perp &= \{x_1, y_1\}, \\ p_1^\perp \cap p_2^\perp \cap \tau(p_3)^\perp &= \{\pi(p_2, p_1 \vee \tau(p_3))\}, \\ p_1^\perp \cap \tau(p_2)^\perp \cap p_3^\perp &= \{x_2, y_2\}, \\ p_1^\perp \cap \tau(p_2)^\perp \cap \tau(p_3)^\perp &= \{\pi(\tau(p_2), p_1 \vee \tau(p_3))\}, \end{aligned}$$

and $c(x_1)$, $c(x_2)$, $c(y_1)$, $c(y_2)$, $c(\pi(p_2, (p_1 \vee \tau(p_3))))$ and $c(\pi(\tau(p_2), (p_1 \vee \tau(p_3))))$ are the six circles touching k_1, k_2 and k_3.

If $|k_1 \cap k_2| = 2$ but $|k_1 \cap k_3| = |k_2 \cap k_3| = 1$ we may assume that p_1 and p_2 are collinear with $\tau(p_3)$, hence p_1, p_2 and p_3 are mutually non-collinear. Thus, there are two points x, y such that:

$$p_1^\perp \cap p_2^\perp \cap p_3^\perp = \{x, y\}.$$

Moreover:

$$\begin{aligned} p_1^\perp \cap p_2^\perp \cap \tau(p_3)^\perp &= \{\pi(p_2, p_1 \vee \tau(p_3))\} = \{\tau(p_3)\}, \\ p_1^\perp \cap \tau(p_2)^\perp \cap p_3^\perp &= \{\pi(p_1, \tau(p_2) \vee p_3)\}, \\ p_1^\perp \cap \tau(p_2)^\perp \cap \tau(p_3)^\perp &= \{\pi(\tau(p_2), p_1 \vee \tau(p_3))\}. \end{aligned}$$

Hence k_3, $c(x)$, $c(y)$, $c(\pi(p_1, \tau(p_2) \vee p_3))$ and $c(\pi(\tau(p_2), p_1 \vee \tau(p_3)))$ are the five circles touching k_1, k_2 and k_3.

If $|k_1 \cap k_2| = |k_1 \cap k_3| = |k_2 \cap k_3| = 1$ we may assume again that p_1 and p_2 are collinear with $\tau(p_3)$, hence p_1, p_2 and p_3 are mutually non-collinear and p_1 is also collinear with $\tau(p_2)$. Let s, t, r such that $k_1 \cap k_2 = \{s\}$, $k_2 \cap k_3 = \{t\}$ and $k_1 \cap k_3 = \{r\}$. Let μ denote the square projection from $Trace_{\tau(p_2)}p_3$ to $g = s \vee p_1$ and let $B := \{\pi(q, g) \mid q \in k_3\}$. Then B is the interval in g with endpoints p_1 and p_2 not containing s and $A := \mu(\tau(p_2)^\perp \cap p_3^\perp)$ is an interval in g with endpoints p_2 and s. By the cases already considered, we know that the interior of B is in A, hence also $B \subset A$. In particular, $p_1 \in A$ and thus there are two points x and y such that:

$$p_1^\perp \cap p_2^\perp \cap p_3^\perp = \{x, y\}$$

and

$$\begin{aligned} p_1^\perp \cap p_2^\perp \cap \tau(p_3)^\perp &= \{\tau(p_3)\}, \\ p_1^\perp \cap \tau(p_2)^\perp \cap p_3^\perp &= \{\tau(p_2)\}, \\ p_1^\perp \cap \tau(p_2)^\perp \cap \tau(p_3)^\perp &= \{p_1\}. \end{aligned}$$

Thus $c(x)$, $c(y)$, k_1, k_2 and k_3 are the five circles touching k_1, k_2 and k_3.

We are left with the cases where $k_1 \cap k_3$ is empty. If $|k_1 \cap k_2| = 1$ and $k_2 \cap k_3 = \emptyset$, then applications of Lemma 7.2 to situations treated above yield the desired results.

For the remaining cases we have to look closer at the involved square projections. Let q denote a cycle such that $|c(q) \cap k_i| = 2$ for $i = 1, 3$. Take $s \in c(q) \cap k_1$, let $g := s \cup p_1$ and consider the square projections $\mu : q^\perp \cap p_3^\perp \to g$ and $\nu : q^\perp \cap \tau(p_3)^\perp \to g$. Then $\mu(q^\perp \cap p_3^\perp) \subset g$ is a closed interval with endpoints s and $\pi(p_3, g)$. Analogously, $\nu(q^\perp \cap \tau(p_3)^\perp) \subset g$ is a closed interval with endpoints s and $\pi(\tau(p_3), g)$. Moreover, let $B := \{\pi(t, g) \mid t \in k_3\}$. As before, B is an interval in g with endpoints $\pi(p_3, g)$ and $\pi(\tau(p_3), g)$. Arguments as above show that B is contained in $\mu(q^\perp \cap p_3^\perp) \cap \nu(q^\perp \cap \tau(p_3)^\perp)$. Thus $B = \mu(q^\perp \cap p_3^\perp) \cap \nu(q^\perp \cap \tau(p_3)^\perp) \setminus \{s\}$ and $g = \mu(q^\perp \cap p_3^\perp) \cup \nu(q^\perp \cap \tau(p_3)^\perp)$. Therefore, if $k_1 \cap k_3 = \emptyset$, i.e., if $p_1 \notin B$, then p_1 is either in $\mu(q^\perp \cap p_3^\perp)$ or in $\nu(q^\perp \cap \tau(p_3)^\perp)$. After possibly exchanging q and $\tau(q)$ in the above argument we may assume $p_1 \in \mu(q^\perp \cap p_3^\perp)$. Hence there are points x_2 and y_2 such that:

$$\begin{aligned} p_1^\perp \cap q^\perp \cap p_3^\perp &= \{x_1, y_1\}, \\ p_1^\perp \cap q^\perp \cap \tau(p_3)^\perp &= \emptyset, \\ p_1^\perp \cap \tau(q)^\perp \cap p_3^\perp &= \{x_2, y_2\}, \\ p_1^\perp \cap \tau(q)^\perp \cap \tau(p_3)^\perp &= \emptyset, \end{aligned}$$

or, equivalently:

$$\begin{aligned} q^\perp \cap p_1^\perp \cap p_3^\perp &= \{x_1, y_1\}, \\ q^\perp \cap p_1^\perp \cap \tau(p_3)^\perp &= \emptyset, \\ q^\perp \cap \tau(p_1)^\perp \cap p_3^\perp &= \emptyset, \\ q^\perp \cap \tau(p_1)^\perp \cap \tau(p_3)^\perp &= \{\tau(x_2), \tau(y_2)\}, \end{aligned}$$

The reason for this reformulation is that now we can shift our point of view. Instead of the square projections μ and ν we now look at the square projections

$$\begin{aligned} &\varphi : p_1^\perp \cap p_3^\perp \to h : x \mapsto \pi(x, h), \\ &\psi : p_1^\perp \cap \tau(p_3)^\perp \to h : x \mapsto \pi(x, h), \\ &\overline{\varphi} : \tau(p_1)^\perp \cap p_3^\perp \to h : x \mapsto \pi(x, h), \\ &\overline{\psi} : \tau(p_1)^\perp \cap \tau(p_3)^\perp \to h : x \mapsto \pi(x, h), \end{aligned}$$

where $h := s \vee q$. Again $\varphi(p_1^\perp \cap p_3^\perp)$ and $\overline{\varphi}(\tau(p_1)^\perp \cap p_3^\perp)$ are closed intervals in h with endpoints s and $\pi(p_3, h)$. Likewise $\psi(p_1^\perp \cap \tau(p_3)^\perp)$ and $\overline{\psi}(\tau(p_1)^\perp \cap \tau(p_3)^\perp)$ are closed intervals in h with endpoints s and $\pi(\tau(p_3), h)$. However, $q \in \varphi(p_1^\perp \cap p_3^\perp)$ and $q \in \overline{\psi}(\tau(p_1)^\perp \cap \tau(p_3)^\perp)$ but $q \notin \overline{\varphi}(\tau(p_1)^\perp \cap p_3^\perp)$ and $q \notin \psi(p_1^\perp \cap \tau(p_3)^\perp)$. Thus $\varphi(p_1^\perp \cap p_3^\perp) \cap \overline{\varphi}(\tau(p_1)^\perp \cap p_3^\perp) = \{s, \pi(p_3, h)\}$ and $\varphi(p_1^\perp \cap p_3^\perp) \cup \overline{\varphi}(\tau(p_1)^\perp \cap p_3^\perp) = h$. Likewise, $\psi(p_1^\perp \cap \tau(p_3)^\perp) \cap \overline{\psi}(\tau(p_1)^\perp \cap \tau(p_3)^\perp) = \{s, \pi(\tau(p_3), h)\}$ and $\psi(p_1^\perp \cap \tau(p_3)^\perp) \cup \overline{\psi}(\tau(p_1)^\perp \cap \tau(p_3)^\perp) = h$.

If we take $p_2 \in h \setminus \{s, \pi(p_3, h), \pi(\tau(p_3), h)\}$, then one of the two sets $\varphi^{-1}(p_2)$ and $\overline{\varphi}^{-1}(p_2)$ contains precisely two points x_1 and y_1 while the other is empty. Also one of the two sets $\psi^{-1}(p_2)$ and $\overline{\psi}^{-1}(p_2)$ contains precisely two points x_2 and y_2 while the other is empty. If $p_2 = \pi(p_3, h)$, i.e., if p_3 is collinear with p_2 and hence $|k_3 \cap k_2| = 1$, then $\varphi^{-1}(p_2) = \pi(p_1, p_2 \vee p_3) =: x_1$ and $\overline{\varphi}^{-1}(p_2) = \pi(\tau(p_1), p_2 \vee p_3) =: y_1$ and again one of the two sets $\psi^{-1}(p_2)$ and $\overline{\psi}^{-1}(p_2)$ contains precisely two points x_2 and y_2 while the other is empty. For $p_2 = \pi(\tau(p_3), h)$ the situation is analogous. So, if $|k_1 \cap k_2| = 2$ and $k_1 \cap k_3 = \emptyset$, then there are always precisely four circles $c(x_1)$, $c(x_2)$, $c(y_1)$ and $c(y_2)$ touching k_1, k_2 and k_3 independent of the number of points in the intersection $k_2 \cap k_3$.

The only case left is the situation where $|k_1 \cap k_2| = |k_2 \cap k_3| = 1$ and $k_1 \cap k_3 = \emptyset$. We may assume that p_2 is collinear with p_1 and p_3. Then $p_1, \tau(p_2)$ and p_3 are mutually non-collinear. Connectedness arguments show that if $k_1 \setminus k_2$ and $k_3 \setminus k_2$ lie in different connected components of $\mathcal{S} \setminus k_2$, then k_2, $c(\pi(\tau(p_3), p_1 \vee p_2))$ and $c(\pi(p_1, \tau(p_3) \vee \tau(p_2)))$ are the only three circles touching k_1, k_2 and k_3.

Now let s denote the element of $k_2 \cap k_3$, let $g := s \vee p_3$ and consider the square projection $\mu : p_1^\perp \cap \tau(p_2)^\perp \to g$. The endpoints of $A := \mu(p_1^\perp \cap \tau(p_2)^\perp)$ are $\pi(\tau(p_2), g) = s$ and $\pi(p_1, g)$. The circle $l := c(\pi(p_1, g))$ separates the tangent pencil $G := c(g)$ into two connected components X and Y. The component X contains all circles of G different from l that meet k_1 or that have no point in common with the connected component of $\mathcal{S} \setminus k_2$ containing $k_1 \setminus k_2$. The component Y contains all circles missing k_1 that are essentially in the same connected component of $\mathcal{S} \setminus k_2$ as $k_1 \setminus k_2$. By the above we have $X \not\subset A$, hence $A = Y \cup \{l, s\}$. Thus, if $k_3 \setminus k_2$ is in the same connected component of $\mathcal{S} \setminus k_2$ as $k_1 \setminus k_2$, i.e., if $k_3 \in Y$, then there are two points x and y such that:

$$p_1^\perp \cap \tau(p_2)^\perp \cap p_3^\perp = \{x, y\}.$$

Moreover:

$$\begin{aligned} p_1^\perp \cap p_2^\perp \cap p_3^\perp &= \{p_2\}, \\ p_1^\perp \cap p_2^\perp \cap \tau(p_3)^\perp &= \{\pi(\tau(p_3), p_1 \vee p_2)\}, \\ p_1^\perp \cap \tau(p_2)^\perp \cap \tau(p_3)^\perp &= \{\pi(p_1, \tau(p_3) \vee \tau(p_2))\}. \end{aligned}$$

Hence $c(x)$, $c(y)$, k_2, $c(\pi(\tau(p_3), p_1 \vee p_2))$ and $c(\pi(p_1, \tau(p_3) \vee \tau(p_2)))$ are the five circles touching k_1, k_2 and k_3.

This completes the proof of the correctness of Table 2.

Remark: A closer look at Table 2 reveals that for every number n between 0 and 8 there is a configuration having n touching circles, except for $n = 7$. For the classical case this was discussed in [32]. Arguments in the corresponding quadrangle show why there never can be a 'seventh circle'.

7.3 The problem of Apollonius in Minkowski planes

We finally come to the problem of Apollonius in Minkowski planes. For locally compact Minkowski planes with parameter 1 the basic strategy is the same as for Möbius planes with parameter 1. There are some additional technical problems due to the existence of parallel classes. The discussion of locally compact Minkowski planes with parameter 2 is comparatively easy, due to the fact that the associated quadrangle is centric. Throughout this section p_i denotes an element of $c^{-1}(k_i)$. Thus $c^{-1}(k_i) = \{p_i, \tau(p_i)\}$ where τ is the covering transformation on $\widetilde{\mathcal{K}}$.

Two points and one circle

The non-trivial cases where two non-parallel points s_1, s_2 and one circle k are given are best treated in the derived projective plane $\mathfrak{P}(\mathfrak{Mi}, s_1)$. There k becomes an oval $\overline{k}$ and Theorem 1.1 can be employed.

If $\{\delta(s_1, s_2), \delta(s_2, s_1)\} \in k$, then in the derived projective plane $\mathfrak{P}(\mathfrak{Mi}, s_1)$ the plus-parallel class and the minus-parallel class through s_2 correspond to two tangents at $\overline{k}$ through s_2. By Theorem 1.1 there is no other line through s_2 tangent to $\overline{k}$, thus there is no circle through s_1 and s_2 touching k.

If $\delta(s_1, s_2) \in k$ and $\delta(s_2, s_1) \notin k$, then in the derived projective plane the parallel class $\Gamma_-(s_2)$ corresponds to a line through s_2 touching $\overline{k}$. So, again by Theorem 1.1, there is one other line through s_2 touching $\overline{k}$. This line corresponds to the unique circle through s_1 and s_2 touching k.

If $\{\delta(s_1, s_2), \delta(s_2, s_1)\} \cap k = \emptyset$ let $t := \delta(\gamma_-(s_1, k), \gamma_+(s_1, k))$. We first consider the parameter 1 case. In the derived projective plane $\mathfrak{P}(\mathfrak{Mi}, s_1)$ the point t is outside $\overline{k}$ since $\Gamma_+(t)$ and $\Gamma_-(t)$ are lines through t tangent to $\overline{k}$. Hence, if $\Gamma(s_1) \cup k$ does not separate s_2 and t, then s_2 is also outside $\overline{k}$ and there are precisely two lines through s_2 tangent to $\overline{k}$. These two lines correspond to the two circles through s_1 and s_2 touching k. If $\Gamma(s_1) \cup k$ separates s_2 and t, then s_2 is inside $\overline{k}$ and no line through s_2 is tangent to $\overline{k}$. Thus no line through s_1 and s_2 touches k. In the parameter 2 case $\mathcal{S} \setminus (\Gamma(s_1) \cup k)$ is connected and in the derived plane there are precisely two lines through s_2 tangent to $\overline{k}$. These two lines correspond to the two circles through s_1 and s_2 touching k.

One point and two circles

Next suppose we are given one point s and two circles k_1 and k_2 contained in a common connected component of $\mathcal{K}$.

If $|k_1 \cap k_2| = 2$ and $k_1 \cap k_2 \cap \Gamma(s) \setminus \{s\} = \{x\}$, then $x \in p_1^\perp \cap p_2^\perp \cap s^\perp$ and $x \in p_1^\perp \cap \tau(p_2)^\perp \cap s^\perp$. Thus, by antiregularity, there are cycles y and z such that $\{x, y\} = p_1^\perp \cap p_2^\perp \cap s^\perp$ and $\{x, z\} = p_1^\perp \cap \tau(p_2)^\perp \cap s^\perp$. Moreover, $\{x, \tau(y)\} = \tau(p_1)^\perp \cap \tau(p_2)^\perp \cap s^\perp$ and $\{x, \tau(z)\} = \tau(p_1)^\perp \cap p_2^\perp \cap s^\perp$. We have $c(y) \neq c(z)$, for otherwise $y \in \{z, \tau(z)\}$. However, if $y = z$, then $y \in p_1^\perp \cap p_2^\perp \cap \tau(p_2)^\perp \cap s^\perp \subset p_2^\perp \cap \tau(p_2)^\perp \subset \mathcal{S}$. But by assumptions x is the only point in $\mathcal{S} \cap p_1^\perp \cap p_2^\perp \cap s^\perp$, hence $y = x$, a contradiction to the choice of y. If $y = \tau(z)$, then $y \in p_1^\perp \cap \tau(p_1)^\perp \cap p_2^\perp \cap s^\perp \subset \mathcal{S}$ which implies again a contradiction. Thus $c(y) = c(\tau(y))$ and $c(z) = c(\tau(z))$ are the two circles through s touching k_1 and k_2.

In the following cases we will leave the proof that the cycles in question are different to the reader, since the argument runs always as above.

If $|k_1 \cap k_2| = 2$ and $|k_1 \cap k_2 \cap \Gamma(s)| = 2$, then $k_1 \cap k_2 = \{\gamma_+(s, k_1), \gamma_-(s, k_1)\}$ and thus, by antiregularity, $\{\gamma_+(s, k_1), \gamma_-(s, k_1)\} = p_1^\perp \cap p_2^\perp \cap s^\perp$. Hence there is no circle through s touching k_1 and k_2

Next suppose $|k_1 \cap k_2| = 2$ and $|k_1 \cap k_2 \cap \Gamma(s)| = 0$ and that s is in one of the components of $\mathcal{S} \setminus (k_1 \cup k_2)$ not containing points parallel to $k_1 \cap k_2$. This configuration can only occur in the parameter 1 case, since in the parameter 2 case the space $\mathcal{S} \setminus (k_1 \cup k_2)$ is connected. Suppose l is a circle through s touching k_1 and k_2. Then $l \cap k_1 \cap k_2$ is empty. On the other hand, $l \setminus (k_1 \cup k_2)$ is contained in one component of $\mathcal{S} \setminus (k_1 \cup k_2)$, namely the one containing s. But then there is no point on l parallel to a point in $k_1 \cap k_2$, which is impossible. Therefore, there exists no circle l through s touching k_1 and k_2. Now let $g = \Gamma_+(s)$ and consider the square projection $\mu : p_1^\perp \cap p_2^\perp \to g : x \mapsto \pi(x, g)$. We are still in the parameter 1 case, so the image $\mu(p_1^\perp \cap p_2^\perp)$ is an interval in g with endpoints $\pi(p_1, g)$ and $\pi(p_2, g)$ (Proposition 2.4). By the above we know that the intersection of $\mu(p_1^\perp \cap p_2^\perp)$ with the component of $\mathcal{S} \setminus (k_1 \cup k_2)$ containing no points parallel to a point in $k_1 \cap k_2$ is empty. Hence $\mu(p_1^\perp \cap p_2^\perp)$ is the closure of the component of $\Gamma_+(s) \setminus (k_1 \cup k_2)$ containing points parallel to a point in the intersection $k_1 \cap k_2$. Thus, if s is in the component of $\mathcal{S} \setminus (k_1 \cup k_2)$ containing points parallel to a point in $k_1 \cap k_2$, then antiregularity implies that the preimage $\mu^{-1}(s)$ contains precisely two points v and w, i.e., $p_1^\perp \cap p_2^\perp \cap s^\perp = \{v, w\}$. Analogous arguments show that there are also two points x and y such that $p_1^\perp \cap \tau(p_2)^\perp \cap s^\perp = \{x, y\}$. In the parameter 2 case we find cycles v, w, x and y with $p_1^\perp \cap p_2^\perp \cap s^\perp = \{v, w\}$ and $p_1^\perp \cap \tau(p_2)^\perp \cap s^\perp = \{x, y\}$ by centric antiregularity. As above one verifies $\{x, y\} \cap \{v, w\} = \emptyset$. Thus $c(v)$, $c(w)$, $c(x)$ and $c(y)$ are the four circles through s touching k_1 and k_2.

If $s \in k_1$ but $s \notin k_2$, then, by Lemma 5.13 and Lemma 5.14, there are precisely two circles touching k_2 that touch k_1 in s. These circles are $c(\pi(p_2, s \vee p_1))$ and $c(\pi(\tau(p_2), s \vee p_1))$.

The configuration $s \notin k_i$ and $k_1 \cap k_2 = \emptyset$ can only occur in the parameter 1 case, since in locally compact Minkowski planes with parameter 2 any two circles intersect. Let $g := \Gamma_+(s)$ and consider the square projections $\mu : p_1^\perp \cap p_2^\perp \to g : x \mapsto \pi(x, g)$ and $\nu : p_1^\perp \cap \tau(p_2)^\perp \to g : x \mapsto \pi(x, g)$. Then $\mu(p_1^\perp \cap p_2^\perp)$ as well as $\nu(p_1^\perp \cap \tau(p_2)^\perp)$ are intervals with endpoints $\pi(p_1, g)$ and $\pi(p_2, g) = \pi(\tau(p_2), g)$. So, either $\mu(p_1^\perp \cap p_2^\perp) \cap \nu(p_1^\perp \cap \tau(p_2)^\perp) = \{\pi(p_1, g), \pi(p_2, g)\}$ and $\mu(p_1^\perp \cap p_2^\perp) \cup \nu(p_1^\perp \cap \tau(p_2)^\perp) = g$ or $\mu(p_1^\perp \cap p_2^\perp) = \nu(p_1^\perp \cap \tau(p_2)^\perp)$. We have to exclude the second case, for in the first case we know that for every $t \in g \setminus \{\pi(p_1, g), \pi(p_2, g)\}$ the union $\mu^{-1}(t) \cup \nu^{-1}(t)$ contains precisely two elements. In particular, we know that there are two points x and y with $\{x, y\} = \mu^{-1}(s) \cup \nu^{-1}(s)$, thus $c(x)$ and $c(y)$ are the two circles through s touching k_1 and k_2. Take $z \in k_1$ not parallel to s. Then, by Lemma 5.13, there are two circles l_1 and l_2 in the touching pencil $B(z, k_1)$ touching k_2. We can choose z such that $l_i \cap k_2$ is not parallel to s for $i = 1, 2$. Considering the configuration in the derivation $\mathfrak{P}(\mathfrak{Mi}, z)$ one sees that $l_1 \setminus (k_1 \cup k_2)$ and $l_2 \setminus (k_1 \cup k_2)$ lie in different connected components of $\mathcal{S} \setminus (k_1 \cup k_2)$, i.e., $l_1 \cap \Gamma_+(s)$ and $l_2 \cap \Gamma_+(s)$ lie in different connected components of $\Gamma_+(s) \setminus (k_1 \cup k_2) = g \setminus \{\pi(p_1, g), \pi(p_2, g)\}$. Let $q_i \in c^{-1}(l_i)$. We may assume $q_1 \in p_1^\perp \cap p_2^\perp$. But then q_2 is either in $\tau(p_1)^\perp \cap p_2^\perp$ or in $p_1^\perp \cap \tau(p_2)^\perp$. Interchanging q_2 with $\tau(q_2)$ if necessary we may assume $q_2 \in p_1^\perp \cap \tau(p_2)^\perp$. Thus $l_1 \cap \Gamma_+(s) = \mu(q_1)$ and $l_2 \cap \Gamma_+(s) = \nu(q_2)$ lie in different connected components of $\Gamma_+(s) \setminus (k_1 \cup k_2)$, hence $\mu(p_1^\perp \cap p_2) \neq \nu(p_1^\perp \cap \tau(p_2)^\perp)$ as claimed.

Three circles

Finally, suppose we are given three circles k_1, k_2 and k_3 all contained in one connected component of $\mathcal{K}$.

The configuration that the circles are pairwise disjoint can only occur in the flat case. Then $\mathcal{S} \setminus (k_1 \cup k_2 \cup k_3)$ splits into three connected components. A circle l touching all three circles has to be contained in the closure of one of these components, but then it can meet only two of the given circles. Thus there is no such circle l.

If there are points y and z such that $k_1 \cap k_2 \cap k_3 = \{y, z\}$, then $p_1^\perp \cap p_2^\perp \cap p_3^\perp = \{y, z\}$ for any choice $p_i \in c^{-1}(k_i)$. Thus there is no circle touching touching k_1, k_2 and k_3.

Next suppose there is a point z such that $k_1 \cap k_2 \cap k_3 = \{z\}$. If $k_1 \cap k_2 = k_1 \cap k_3 = k_2 \cap k_3 = \{z\}$ then there are infinitely many circles touching the three given circles, namely all the circles contained in the touching pencil spanned by these three circles.

The case $|k_1 \cap k_2| = 2$ and $|k_1 \cap k_3| = |k_2 \cap k_3| = 1$ cannot occur, since in this situation we may assume that p_3 is collinear with p_1 and p_2. But z is collinear with p_1, p_2 and p_3 which is only possible if z and p_3 are on a line spanned by p_1 and p_2, i.e., if p_1 is collinear with p_2. This implies $k_1 \cap k_2 = \{z\}$, contradicting our assumptions.

If $|k_1 \cap k_2| = |k_2 \cap k_3| = 2$ and $|k_1 \cap k_3| = 1$ we may assume that p_1 is collinear with p_3. This implies that for $q \in \{p_2, \tau(p_2)\}$ we have $p_1^\perp \cap q^\perp \cap p_3^\perp = \{\pi(q, p_1 \vee p_3)\} = \{z\}$. However, if p_1 is collinear with p_3, then p_1 is not collinear with $\tau(p_3)$, thus, by antiregularity, there are cycles x and y such that $p_1^\perp \cap p_2^\perp \cap \tau(p_3)^\perp = \{x, z\}$ and $p_1^\perp \cap \tau(p_2)^\perp \cap \tau(p_3)^\perp = \{y, z\}$. Then $c(x)$ and $c(y)$ are the two circles touching k_1, k_2 and k_3. If, still under the assumption that there is a point z such that $k_1 \cap k_2 \cap k_3 = \{z\}$, we have $|k_i \cap k_j| = 2$ for $i \neq j$, then $z \in p_1^\perp \cap p_2^\perp \cap p_3^\perp \cap \tau(p_2)^\perp \cap \tau(p_3)^\perp$. By antiregularity there are cycles v, w, x and y such that:

$$\begin{aligned} p_1^\perp \cap p_2^\perp \cap p_3^\perp &= \{v, z\}, \\ p_1^\perp \cap p_2^\perp \cap \tau(p_3)^\perp &= \{w, z\}, \\ p_1^\perp \cap \tau(p_2)^\perp \cap p_3^\perp &= \{x, z\}, \\ p_1^\perp \cap \tau(p_2)^\perp \cap \tau(p_3)^\perp &= \{y, z\}. \end{aligned}$$

Thus $c(v)$, $c(w)$, $c(x)$ and $c(y)$ are the four circles touching k_1, k_2 and k_3.

We are now left with the cases where k_1 and k_2 have at least one point in common, but there is no point incident with all three circles. We first restrict our attention to the parameter 1 case. Here square projections will be our main tool. So let's have a closer look at this map in the situation where there is a point $s \in k_1 \cap k_2 \setminus k_3$ and p_1, p_2 and p_3 are pairwise non-collinear. Let $g := s \vee p_1$ denote the line connecting s to p_1 and consider the square projection $\mu : p_2^\perp \cap p_3^\perp \to g : x \mapsto \pi(x, g)$. Again, $\mu(p_2^\perp \cap p_3^\perp)$ is a closed interval in g with endpoints $\pi(p_2, g) = s$ and $\pi(p_3, g)$. Let A denote a connected component of $k_3 \setminus \Gamma(s)$. Then $\pi(A, g)$ is either contained entirely in $\mu(p_2^\perp \cap p_3^\perp)$ or contained entirely in the closure of the complement of $\mu(p_2^\perp \cap p_3^\perp)$ in g.

Now suppose the intersection $k_2 \cap k_3$ is not empty and let $t \in k_2 \cap k_3$. Then $t \in p_2^\perp \cap p_3^\perp$, hence $\pi(t, g) = \mu(t) \in \mu(p_2^\perp \cap p_3^\perp)$. Therefore, $\mu(p_2^\perp \cap p_3^\perp)$ contains $\pi(A, g)$ if and only if A is the connected component of $k_3 \setminus \Gamma(s)$ containing $k_3 \cap k_2$. Thus, if $k_1 \cap k_3 \neq \emptyset$ is in the same connected component of $k_3 \setminus \Gamma(s)$ as $k_2 \cap k_3$, then $p_1 \in \mu(p_2^\perp \cap p_3^\perp)$ and hence $p_1^\perp \cap p_2^\perp \cap p_3^\perp = \mu^{-1}(p_1)$ contains precisely two points. Conversely, if $k_1 \cap k_3 \neq \emptyset$ is not in the same connected component of $k_3 \setminus \Gamma(s)$

as $k_2 \cap k_3$, then $p_1 \notin \mu(p_2^\perp \cap p_3^\perp)$ and hence $p_1^\perp \cap p_2^\perp \cap p_3^\perp = \mu^{-1}(p_1) = \emptyset$. This result allows us to treat all the cases still left where any two of the circles k_1, k_2 and k_3 have at least one point in common.

If $|k_1 \cap k_2| = |k_1 \cap k_3| = |k_2 \cap k_3| = 2$, then in the above argument we can replace any of the points p_i by $\tau(p_i)$. Thus, if $k_1 \cap k_3$ is in the same connected component of $k_3 \setminus \Gamma(s)$ as $k_2 \cap k_3$, there are eight points x_1, x_2, x_3, x_4, y_1, y_2, y_3 and y_4 such that:

$$\begin{aligned} p_1^\perp \cap p_2^\perp \cap p_3^\perp &= \{x_1, y_1\}, \\ p_1^\perp \cap p_2^\perp \cap \tau(p_3)^\perp &= \{x_2, y_2\}, \\ p_1^\perp \cap \tau(p_2)^\perp \cap p_3^\perp &= \{x_3, y_3\}, \\ p_1^\perp \cap \tau(p_2)^\perp \cap \tau(p_3)^\perp &= \{x_4, y_4\}. \end{aligned}$$

Then $c(x_1)$, $c(x_2)$, $c(x_3)$, $c(x_4)$, $c(y_1)$, $c(y_2)$, $c(y_3)$ and $c(y_4)$ are the eight circles touching k_1, k_2 and k_3. Conversely, if $k_1 \cap k_3$ is not in the same connected component of $k_3 \setminus \Gamma(s)$ as $k_2 \cap k_3$, then for any choice of $q_i \in \{p_i, \tau(p_i)\}$ the intersection $q_1^\perp \cap q_2^\perp \cap q_3^\perp$ is empty, thus there are no circles touching k_1, k_2 and k_3.

If $|k_1 \cap k_2| = |k_2 \cap k_3| = 2$ but $|k_1 \cap k_3| = 1$, then we may assume that p_1 is collinear with $\tau(p_3)$, hence not with p_3. This implies that in the above argument we can only replace p_2 by $\tau(p_2)$ or p_1 and p_3 by $\tau(p_1)$ and $\tau(p_3)$ simultaneously. Thus, if $k_1 \cap k_3$ is in the same connected component of $k_3 \setminus \Gamma(s)$ as $k_2 \cap k_3$, then there are four points x_1, x_2, y_1 and y_2 such that:

$$\begin{aligned} p_1^\perp \cap p_2^\perp \cap p_3^\perp &= \{x_1, y_1\}, \\ p_1^\perp \cap p_2^\perp \cap \tau(p_3)^\perp &= \{\pi(p_2, p_1 \vee \tau(p_3))\}, \\ p_1^\perp \cap \tau(p_2)^\perp \cap p_3^\perp &= \{x_2, y_2\}, \\ p_1^\perp \cap \tau(p_2)^\perp \cap \tau(p_3)^\perp &= \{\pi(\tau(p_2), p_1 \vee \tau(p_3))\}. \end{aligned}$$

Thus $c(x_1)$, $c(x_2)$, $c(y_1)$ and $c(y_2)$, $c(\pi(p_2, (p_1 \vee \tau(p_3))))$ and $c(\pi(\tau(p_2), (p_1 \vee \tau(p_3))))$ are the six circles touching k_1, k_2 and k_3. Conversely, if $k_1 \cap k_3$ is not in the same connected component of $k_3 \setminus \Gamma(s)$ as $k_2 \cap k_3$, then for any choice of $q \in \{p_2, \tau(p_2)\}$ the intersection $q_1^\perp \cap q^\perp \cap q_3^\perp$ is empty, thus $c(\pi(p_2, (p_1 \vee \tau(p_3))))$ and $c(\pi(\tau(p_2), (p_1 \vee \tau(p_3))))$ are the two circles touching k_1, k_2 and k_3.

If $|k_1 \cap k_2| = 2$ but $|k_1 \cap k_3| = |k_2 \cap k_3| = 1$ we may assume that p_1 and p_2 are collinear with $\tau(p_3)$, hence p_1, p_2 and p_3 are mutually non-collinear. Thus, if

$k_1 \cap k_3$ is in the same connected component of $k_3 \setminus \Gamma(s)$ as $k_2 \cap k_3$, then there are two points x, y such that:

$$p_1^\perp \cap p_2^\perp \cap p_3^\perp = \{x, y\}.$$

Moreover:

$$\begin{aligned} p_1^\perp \cap p_2^\perp \cap \tau(p_3)^\perp &= \{\pi(p_2, p_1 \vee \tau(p_3))\} = \{\tau(p_3)\}, \\ p_1^\perp \cap \tau(p_2)^\perp \cap p_3^\perp &= \{\pi(p_1, \tau(p_2) \vee p_3)\}, \\ p_1^\perp \cap \tau(p_2)^\perp \cap \tau(p_3)^\perp &= \{\pi(\tau(p_2), p_1 \vee \tau(p_3))\}. \end{aligned}$$

Thus k_3, $c(x)$, $c(y)$, $c(\pi(p_1, \tau(p_2) \vee p_3))$ and $c(\pi(\tau(p_2), p_1 \vee \tau(p_3)))$ are the five circles touching k_1, k_2 and k_3. Conversely, if $k_1 \cap k_3$ is not in the same connected component of $k_3 \setminus \Gamma(s)$ as $k_2 \cap k_3$, then the intersection $q_1^\perp \cap q^\perp \cap q_3^\perp$ is empty, thus k_3, $c(\pi(p_1, (\tau(p_2) \vee p_3)))$ and $c(\pi(\tau(p_2), (p_1 \vee \tau(p_3))))$ are the three circles touching k_1, k_2 and k_3.

If $|k_1 \cap k_2| = |k_1 \cap k_3| = |k_2 \cap k_3| = 1$ we may assume again that p_1 and p_2 are collinear with $\tau(p_3)$, hence p_1, p_2 and p_3 are mutually non-collinear and p_1 is also collinear with $\tau(p_2)$. By Lemma 5.15 we know that $k_1 \cap k_3$ and $k_2 \cap k_3$ are in different connected components of $k_3 \setminus \Gamma(s)$. Thus the centre $p_1^\perp \cap p_2^\perp \cap p_3^\perp$ is empty. However:

$$\begin{aligned} p_1^\perp \cap p_2^\perp \cap \tau(p_3)^\perp &= \{\tau(p_3)\}, \\ p_1^\perp \cap \tau(p_2)^\perp \cap p_3^\perp &= \{\tau(p_2)\}, \\ p_1^\perp \cap \tau(p_2)^\perp \cap \tau(p_3)^\perp &= \{p_1\}. \end{aligned}$$

Thus k_1, k_2 and k_3 are the three circles touching k_1, k_2 and k_3.

For the remaining cases, where we may assume that $k_1 \cap k_3$ is empty, we have to look closer at the involved square projections. Let q denote a cycle such that $|c(q) \cap k_i| = 2$ for $i = 1, 3$. Take $s \in c(q) \cap k_1$, let $g := s \cup p_1$ and consider the square projections $\mu : q^\perp \cap p_3^\perp \to g$ and $\nu : q^\perp \cap \tau(p_3)^\perp \to g$. Then $\mu(q^\perp \cap p_3^\perp) \subset g$ is a closed interval with endpoints s and $\pi(p_3, g)$. Analogously, $\nu(q^\perp \cap \tau(p_3)^\perp) \subset g$ is a closed interval with endpoints s and $\pi(\tau(p_3), g)$. Moreover, if A denotes the connected component of $k_3 \setminus \Gamma(s)$ containing $k_2 \cap k_3$, then $\pi(A, g) \subset \mu(q^\perp \cap p_3^\perp) \cap \nu(q^\perp \cap \tau(p_3)^\perp)$. Thus, after possibly interchanging p_3 and $\tau(p_3)$, we may assume $\pi(A, g) \cup \{s\} = \nu(q^\perp \cap \tau(p_3)^\perp) \subset \mu(q^\perp \cap p_3^\perp)$. Furthermore, if $B := k_3 \setminus (\Gamma(s) \cup A)$ denotes the connected component of $k_3 \setminus \Gamma(s)$ not containing $k_3 \cap k_2$, then $\pi(B, g) \cap \mu(q^\perp \cap p_3^\perp) = \{\pi(p_3, g)\}$. Thus $\mu(q^\perp \cap p_3^\perp) \setminus \nu(q^\perp \cap \tau(p_3)^\perp)$ contains besides the cycle $\pi(p_3, g)$ precisely those cycles v in g for which $c(v) \cap k_3 = \emptyset$. Hence $\nu^{-1}(p_1) = \emptyset$ and there are points x_1 and y_1 such that $\mu^{-1}(p_1) = \{x_1, y_1\}$. We can exchange q by $\tau(q)$ in the above argument and obtain that there are points x_2 and y_2 such that:

$$\begin{aligned} p_1^\perp \cap q^\perp \cap p_3^\perp &= \{x_1, y_1\}, \\ p_1^\perp \cap q^\perp \cap \tau(p_3)^\perp &= \emptyset, \\ p_1^\perp \cap \tau(q)^\perp \cap p_3^\perp &= \{x_2, y_2\}, \\ p_1^\perp \cap \tau(q)^\perp \cap \tau(p_3)^\perp &= \emptyset, \end{aligned}$$

or, equivalently:

$$\begin{aligned} q^\perp \cap p_1^\perp \cap p_3^\perp &= \{x_1, y_1\}, \\ q^\perp \cap p_1^\perp \cap \tau(p_3)^\perp &= \emptyset, \\ q^\perp \cap \tau(p_1)^\perp \cap p_3^\perp &= \emptyset, \\ q^\perp \cap \tau(p_1)^\perp \cap \tau(p_3)^\perp &= \{\tau(x_2), \tau(y_2)\}. \end{aligned}$$

The reason for this reformulation is that now we can shift our point of view. Instead of the square projections μ and ν we now look at the square projections

$$\begin{aligned} \varphi &: p_1^\perp \cap p_3^\perp \to h : x \mapsto \pi(x, h), \\ \psi &: p_1^\perp \cap \tau(p_3)^\perp \to h : x \mapsto \pi(x, h), \\ \overline{\varphi} &: \tau(p_1)^\perp \cap p_3^\perp \to h : x \mapsto \pi(x, h), \\ \overline{\psi} &: \tau(p_1)^\perp \cap \tau(p_3)^\perp \to h : x \mapsto \pi(x, h), \end{aligned}$$

where $h := s \vee q$. Again $\varphi(p_1^\perp \cap p_3^\perp)$ and $\overline{\varphi}(\tau(p_1)^\perp \cap p_3^\perp)$ are closed intervals in h with endpoints s and $\pi(p_3, h)$. Likewise, $\psi(p_1^\perp \cap \tau(p_3)^\perp)$ and $\overline{\psi}(\tau(p_1)^\perp \cap \tau(p_3)^\perp)$ are closed intervals in h with endpoints s and $\pi(\tau(p_3), h)$. However, $q \in \varphi(p_1^\perp \cap p_3^\perp)$ and $q \in \overline{\psi}(\tau(p_1)^\perp \cap \tau(p_3)^\perp)$ but $q \notin \overline{\varphi}(\tau(p_1)^\perp \cap p_3^\perp)$ and $q \notin \psi(p_1^\perp \cap \tau(p_3)^\perp)$. Thus $\varphi(p_1^\perp \cap p_3^\perp) \cap \overline{\varphi}(\tau(p_1)^\perp \cap p_3^\perp) = \{s, \pi(p_3, h)\}$ and $\varphi(p_1^\perp \cap p_3^\perp) \cup \overline{\varphi}(\tau(p_1)^\perp \cap p_3^\perp) = h$. Likewise, $\psi(p_1^\perp \cap \tau(p_3)^\perp) \cap \overline{\psi}(\tau(p_1)^\perp \cap \tau(p_3)^\perp) = \{s, \pi(\tau(p_3), h)\}$ and $\psi(p_1^\perp \cap \tau(p_3)^\perp) \cup \overline{\psi}(\tau(p_1)^\perp \cap \tau(p_3)^\perp) = h$.

Thus, if we take $p_2 \in h \backslash \{s, \pi(p_3, h), \pi(\tau(p_3), h)\}$, then one of the two sets $\varphi^{-1}(p_2)$ and $\overline{\varphi}^{-1}(p_2)$ contains precisely two points x_1 and y_1 while the other is empty. Also, one of the two sets $\psi^{-1}(p_2)$ and $\overline{\psi}^{-1}(p_2)$ contains precisely two points x_2 and y_2 while the other is empty. If $p_2 = \pi(p_3, h)$, i.e., if p_3 is collinear with p_2 and hence $|k_3 \cap k_2| = 1$, then $\varphi^{-1}(p_2) = \pi(p_1, p_2 \vee p_3) =: x_1$ and $\overline{\varphi}^{-1}(p_2) = \pi(\tau(p_1), p_2 \vee p_3) =: y_1$ and again one of the two sets $\psi^{-1}(p_2)$ and $\overline{\psi}^{-1}(p_2)$ contains precisely two points x_2 and y_2 while the other is empty. For $p_2 = \pi(\tau(p_3), h)$ the situation is analogous. So, if $|k_1 \cap k_2| = 2$ and $k_1 \cap k_3 = \emptyset$, then there are always precisely four circles $c(x_1)$,

$c(x_2)$, $c(y_1)$ and $c(y_2)$ touching k_1, k_2 and k_3 independent of the number of points in the intersection $k_2 \cap k_3$.

The only parameter 1 case left is the situation where $|k_1 \cap k_2| = |k_2 \cap k_3| = 1$ and $k_1 \cap k_3 = \emptyset$. We may assume that p_2 is collinear with p_1 and p_3. Then $p_1, \tau(p_2)$ and p_3 are mutually non-collinear. Suppose there is some point $q \in p_1^\perp \cap \tau(p_2)^\perp \cap p_3^\perp$. Then $|c(q) \cap k_1| = |c(q) \cap k_2| = |k_1 \cap k_2| = 1 \neq |k_1 \cap k_2 \cap c(q)|$ and k_3 is a circle touching $c(q)$, k_1 and k_3. This contradicts one of the cases already treated, thus there is no such q and therefore we obtain:

$$\begin{aligned}
p_1^\perp \cap p_2^\perp \cap p_3^\perp &= \{p_2\},\\
p_1^\perp \cap p_2^\perp \cap \tau(p_3)^\perp &= \{\pi(\tau(p_3), p_1 \vee p_2)\},\\
p_1^\perp \cap \tau(p_2)^\perp \cap p_3^\perp &= \emptyset,\\
p_1^\perp \cap \tau(p_2)^\perp \cap \tau(p_3)^\perp &= \{\pi(p_1, \tau(p_3) \vee \tau(p_2))\}.
\end{aligned}$$

Hence k_2, $c(\pi(\tau(p_3), p_1 \vee p_2))$ and $c(\pi(p_1, \tau(p_3) \vee \tau(p_2)))$ are the three circles touching k_1, k_2 and k_3.

In a locally compact Minkowski plane with parameter 2 any two circles meet and a punctured circle is still connected. This limits the number of possible configurations to the ones listed in Table 3a and Table 3b. Moreover, the number of common touching circles follows readily from centric antiregularity. The only interesting configuration left is the situation where $|k_1 \cap k_2| = |k_2 \cap k_3| = |k_2 \cap k_3| = 1$ and $k_1 \cap k_2 \cap k_3 = \emptyset$. As in the parameter 1 case we may assume that p_1, p_2 and p_3 are mutually non-collinear. By centric antiregularity there are two points x and y such that $p_1^\perp \cap p_2^\perp \cap p_3^\perp = \{x, y\}$. Hence $c(x)$, $c(y)$, k_1, k_2 and k_3 are the five circles touching k_1, k_2 and k_3. That this differs from the result in the parameter 1 case has its reason in the difference of the underlying topological configuration. In the parameter 1 case we have seen that $k_1 \cap k_3$ and $k_2 \cap k_3$ lie in different connected components of $k_3 \setminus \Gamma(k_1 \cap k_3)$. In the parameter 2 case however, the set $k_3 \setminus \Gamma(k_1 \cap k_3)$ is connected, thus it corresponds to the impossible parameter 1 situation where $k_1 \cap k_3$ and $k_2 \cap k_3$ lie in the same connected component of $k_3 \setminus \Gamma(k_1 \cap k_3)$.

This shows correctness of Table 3a and Table 3b and thus completes the proof of Theorem 7.1.

Index

Glossary

$\mathbb{N}$	natural numbers	
$\mathbb{R}$	real numbers	
$\mathbb{C}$	complex numbers	
$\mathbb{Z}$	integers and infinite cyclic group	
$\mathbb{S}_n$	n-dimensional sphere	
$\mathbb{D}_n$	n-dimensional ball	
$\mathfrak{Cp}$	circle plane	7
$\mathfrak{L}$	Laguerre plane	9
$\mathfrak{Mo}$	Möbius plane	9
$\mathfrak{Mi}$	Minkowski plane	9
$\mathfrak{GQ}$	generalised quadrangle	16
$\mathcal{S}$	point set of a circle plane	7
$\mathcal{K}$	circle set of a circle plane	7
$\mathcal{E}$	set of equivalence relations	7
$\mathcal{F}$	set of flags	8
$\mathcal{B}$	set of touching pencils	8
$\overline{\mathcal{B}}$	set of extended touching pencils	8
$\mathcal{G}$	$\{\Gamma(s) \mid s \in \mathcal{S}\}$	8
$\mathcal{G}_+$	set of plus-parallel classes	9
$\mathcal{G}_-$	set of minus-parallel classes	9
$\overline{\mathcal{G}}$	set of extended equivalence classes	40
$\widetilde{\mathcal{K}}$	space of cycles	56
$\widetilde{\mathcal{B}}$	space of cycle pencils	56
$\widetilde{\overline{\mathcal{B}}}$	space of extended cycle pencils	56
$\widetilde{\mathcal{F}}$	space of flags in the lifted geometry	56
$\mathcal{K}_s$	set of circles incident with s	8
$\mathcal{K}_{s,t}$	set of circles incident with s and t	8
$\mathcal{K}_i$	connected component of $\mathcal{K}$	76
$\mathcal{P}$	point set of a generalised quadrangle	16

$\mathcal{L}$	line set of a generalised quadrangle	16
$\mathcal{L}_p$	line pencil through p	17
$\mathcal{P}_l$	point row of l	17
$\Delta_i(v)$	vertices at distance i from v	44
$\mathcal{T}$	space of triads	17
$\Gamma(s)$	set of points parallel to s	8
$B(s,k)$	touching pencil spanned by s and k	8
$\overline{B}(s,k)$	extended touching pencil spanned by s and k	8
$\Gamma_+(s)$	plus-parallel class of s	9
$\Gamma_-(s)$	minus-parallel class of s	9
l_∞	line at infinity	14
$p^\perp$	set of points collinear with p	17
$Tr(A)$	set of triples of A	74
$\alpha(s,t,u)$	circle through s, t and u	7
$\beta(s,k,t)$	circle through t touching k in s	7
$\gamma_i(s,k)$	point on k that is i-parallel to s	7
$\delta_{i,j}(s,t)$	point i-parallel to s and j-parallel to t	8
$\varepsilon(t,b)$	circle in b incident with t	80
$\sigma(b)$	support of b	8
$\parallel_+$	plus-parallel	9
$\parallel_-$	minus-parallel	9
$\gamma_+(s,k)$	point on k plus-parallel to s	9
$\gamma_-(s,k)$	point on k minus-parallel to s	9
$\delta(s,t)$	point plus-parallel to s and minus-parallel to t	9
$\parallel$	parallelism in Laguerre planes	9
$\gamma(s,k)$	point on k parallel to s	9
$\pi(p,g)$	point on g collinear with p	16
$\lambda(p,g)$	line through p intersecting g	16
$x \sim y$	x is collinear with y	17
$p \vee q$	line joining p to q	16
$\mu_{p,q,g}$	square projection from $p^\perp \cap q^\perp$ to g	19
$\deg(f)$	mapping degree of f	20
c	covering	56
H^-	hemisphere opposite H	23
$m(\mathfrak{G}\mathfrak{Q})$	twisting number of $\mathfrak{G}\mathfrak{Q}$	25

References

[1] Armstrong, M.A., *Basic Topology*, Springer-Verlag, New York, 1983

[2] Benz, W., *Vorlesung über Geometrie der Algebren*, Die Grundlehren der mathematischen Wissenschaften in Einzeldarstellungen **197**, Springer, Berlin, 1937

[3] Brouwer, L.E.J., *Über die periodischen Transformationen der Kugel*, Math. Ann. **80** (1919), 39–41

[4] Buchanan, T., Hähl, H. and Löwen, R., *Topologische Ovale*, Geom. Dedicata **9** (1980), 401–424

[5] Bruen, A., Fisher, J.C. and Wilker, J.B., *Apollonius by Inversion*, Mathematics Magazine **56** (1983), 97–103

[6] Buekenhout, F., *Le plans de Benz: Une approach unifiee des plans de Moebius, Laguerre et Minkowski*, J. Geom. **17** (1981), 61–68

[7] Coxeter, H.S.M., *The problem of Apollonius*, Amer. Math. Month. **75** (1968), 5–15

[8] Dugundji, J., *Topology*, Allyn and Bacon, Inc., Boston, 1965

[9] Engelking, R., *General Topology*, Heldermann, Berlin, 1989

[10] Fisher, Ch., *Models and Theorems of the Classical Circle planes*, Abh. Math. Sem. Univ. Hamburg **63** (1993), 245–264

[11] Forst, M., *Topologische 4-Gone*, Mitt. d. Math. Sem. Gießen **147** (1981), 65–129

[12] Groh, H., *Topologische Laguerre-Ebenen I*, Abh. Math. Sem. Univ. Hamburg **32** (1968), 216–231

[13] Groh, H., *Topologische Laguerre-Ebenen II*, Abh. Math. Sem. Univ. Hamburg **34** (1969), 11–21

[14] Groh, H., *On flat Laguerre planes*, J. Geometry **1** (1971), 18–40

[15] Groh, H., *Laguerre planes generated by Moebius planes*, Abh. Math. Sem. Univ. Hamburg **40** (1973), 43–63

[16] Groh, H., *Flat Moebius planes and Laguerre planes*, Abh. Math. Sem. Univ. Hamburg **40** (1973), 64–76

[17] Groh, H., *Flat Moebius planes*, Geom. Dedicata **1** (1972), 65–84

[18] Grundhöfer, Th. and Knarr, N., *Topology in generalized quadrangles*, Topology Appl. **34** (1990), 139–152

[19] Heath, Th.L., *A Manual of Greek Mathematics*, Dover Publ., New York, 1963

[20] Hilgert, J. and Hofmann, K.H., *Old and new on Sl(2)*, manuscripta math. **45** (1985), 17–52

[21] Kerékjartó B. von, *Über die periodischen Transformationsgruppen der Kreisscheibe und der Kugelfläche*, Math. Ann. **80** (1919), 36–38

[22] Knarr, N., *The nonexistence of certain topological polygons*, Forum Math. **2** (1990), 603–612

[23] Laguerre, E., *La Théorie des Foyers*, Nouvelles Annales de Mathematiques (1853),

[24] Kramer, L., *Compact Polygons*, Dissertation, Tübingen, 1994

[25] Leißner, W., *Büschelhomogene Lie-Ebenen*, J. Reine Angew. Math. **246** (1971), 76–166

[26] Löwen, R., *Projectivities and the geometric structure of topological planes*, in: Plauman P. and Strambach K.:, Geometry — von Staudt's point of view, Bad Windsheim, 1980, (D. Reidel 1981), 339–372

[27] Löwen, R. and Pfüller, U., *Two-Dimensional Laguerre Planes over Convex Functions*, Geom. Dedicata **23** (1987), 73–85

[28] Löwen, R. and Pfüller, U., *Two-Dimensional Laguerre Planes with Large Automorphism Groups*, Geom. Dedicata **23** (1987), 87–96

[29] Nadler, S.B., *Hyperspaces of sets*, Monographs and textbooks in pure and applied mathematics **49**, Marcel Dekker, New York, 1978

[30] Payne, S.E. and Thas, J.A., *Generalized quadrangles with symmetry*, Simon Stevin **49** (1970), 3–32 and 81–103

[31] Payne, S.E. and Thas, J.A., *Finite Generalized Quadrangles*, Research Notes in Math. **110**, Pitman, Boston, 1984

[32] Pedoe, D., *The Missing seventh circle*, Elem. Math. **25** (1970), 14–15

[33] Pedoe, D., *A Course of Geometry for Colleges and Universities*, Cambridge University Press, Cambridge, 1970

[34] Rigby, J.F., *The geometry of cycles and generalized Laguerre inversions*, The Geometric Vein - The Coxeter Festschrift, 355–378, Springer, 1981.

[35] Schenkel, A., *Topologische Minkowskiebenen*, Dissertation, Erlangen-Nürnberg, 1982

[36] Schroth, A.E., *Dreidimensionale Vierecke und ebene Laguerre-Ebenen*, Diplomarbeit, Tübingen, 1988

[37] Schroth, A.E., *Three Dimensional Quadrangles and Flat Laguerre Planes*, Geom. Dedicata **36** (1990), 365–373

[38] Schroth, A.E., *The Apollonius Problem in Flat Laguerre Planes*, J. Geom. **42** (1991), 141–147

[39] Schroth, A.E., *Topologische Laguerreebenen und topologische Vierecke*, Dissertation, Braunschweig, 1992

[40] Schroth, A.E., *On the structure of topological n-gons*, Simon Stevin **67** (1993), 259–266

[41] Schroth, A.E., *Generalized quadrangles constructed from topological Laguerre planes*, Geom. Dedicata **46** (1993), 339–361

[42] Schroth, A.E., *The Apollonius Problem in four-dimensional Laguerre Planes*, J. Geom. **51** (1994), 138–147

[43] Schroth, A.E., *Topological antiregular Quadrangles*, Resultate der Mathematik **24** (1993), 180–189

[44] Schroth, A.E., *Sisterhoods of flat Laguerre planes*, to appear in Geom. Dedicata

[45] Schroth, A.E. and Van Maldeghem H., *Half regular and regular points in compact polygons*, Geom. Dedicata **51** (1994), 215–233

[46] Schroth, A.E., *Topological Circle Planes and Topological Quadrangles*, Habilitationsschrift, Braunschweig, 1994

[47] Steenrod, N., *The topology of fibre bundles*, Princton University press, Princton, 1951

[48] Steinke, G.F., *Eine Klassifikation 4-dimensionaler Laguerre-Ebenen mit großer Automorphismengruppe*, Habilitationsschrift, Kiel, 1987

[49] Steinke, G.F., *Topological properties of locally compact connected Minkowski planes and their derived affine planes*, Geom. Dedicata **32** (1989), 341–351

[50] Steinke, G.F., *Topological Circle Geometries*, In: Buekenhout, Francis (ed.) Handbook of incidence geometry: buildings and foundations. Amsterdam: North-Holland, 1325–1354, 1995.

[51] Thas, J.A., *Circle geometries and generalized quadrangles*, Finite Geometries, (Winnipeg, Man, 1984), Lecture Notes in Pure and Appl. Math., **103**, 327–352, Dekker, New York, 1985

[52] Thas, J.A., *Generalized Polygons*, In: Buekenhout, Francis (ed.) Handbook of incidence geometry: buildings and foundations. Amsterdam: North-Holland, 383–432, 1995.

[53] Tits, J., *Classification of Buildings of spherical Type and Moufang Polygons: A survey*, Atti. Coll. Intern. Teorie Combinatorie, Accad. Naz. Lincei, 1, Rome, 1973, (1976), 229–246

[54] Tits, J., *Sur la trialité et certains groupes qui s'en déduisent*, Publ. Math. I.H.E.S. **2** (1959), 13–60

[55] Wölk, R., *Topologische Möbiusebenen*, Math. Z. **93** (1966), 311–333